Railwayman -
Engineer – Diver

Memories of a lifelong railway engineer

Brian Maddison

Published by New Generation Publishing in 2024

Paperback ISBN: 978-1-83563-287-1
eBook ISBN: 978-1-83563-288-8

www.newgeneration-publishing.com

New Generation Publishing

Any profits that I receive from the sale of this book will be donated to the Railway Children Charity.   Brian Maddison

Railway Children believe that no child should have to live on the streets. They work across India, Tanzania and the UK to protect children at risk of a life on the streets, where they suffer abuse and exploitation.

Children run away or are forced to leave homes where they suffer poverty, violence, abuse and neglect. They find themselves living on the streets because there is nowhere else to go, and nobody left to turn to. But often the danger they face on the streets is much worse.

Railway Children work every day to ensure that they reach children as soon as they arrive on the streets and get to them before an abuser can. So that together we can give every child a safer, brighter future.

To find out more about Railway Children's work visit www.railwaychildren.org.uk

children RAILWAY

No child lost to the streets

CONTENTS

Chapter 1 Early Days.. 1

Chapter 2 Peterborough – Part 1 11

Chapter 3 Woolsthorpe Branch – Bridge 2 26

Chapter 4 Down to "The Smoke"... 31

Chapter 5 Steam Breakdown Cranes and the Bridge at Harringay........... 36

Chapter 6 The Leicester Experience 46

Chapter 7 Nottingham, Furlong House Drawing Office........................... 56

Chapter 8 Drinking 77

Chapter 9 Lady Elton's Personal Engineer 83

Chapter 10 A move into Management 89

Chapter 11 Tunnel Work and the 100 MPH Tunnel Train 99

Chapter 12 Bridge 79, Loughborough 111

Chapter 13 More Bridge Reconstructions 123

Chapter 14 Return to Peterborough 154

Chapter 15 Permanent Way Stuff and on-call at Peterborough 172

Chapter 16 New Barnet station fire 183

Chapter 17 Chartered Engineer 188

Chapter 18 Works Assistant for Inter-City................................. 194

Chapter 19 Diving – The Early Days 202

Chapter 20 Diving – I'm in Charge 224

Chapter 21 The Diving gets Harder, and we move into the Private Sector
.... 244

Chapter 22 Bridgeway Beginnings... 263

Chapter 23 Possession Management 275

Chapter 24 Money Matters... 284

Chapter 25 What else we did at Bridgeway............................. 288

Chronology

1946   Born in Old Fletton, near Peterborough

1951–1957 Attended Old Fletton Infants and Junior Schools

1957–1962 Attended Old Fletton Grammar School / Orton Longueville Grammar School

1962–1963 Various Jobs including working for Hawkins and Sons, Builders. Also studied Building Construction on day release.

1964–1965 B.R. District Engineer's Office – Peterborough

1965–1967 Divisional Civil Engineer's Office – Kings Cross

1967–1971 Area Civil Engineer's Office, Leicester (Senior Technical Officer)

1971–1974 Divisional Civil Engineer's Office – Nottingham. Building Section

1974–1984 Divisional Civil Engineer's Office – Nottingham. Bridge Section (Principal Technical Office)

1984–1989 DCE Nottingham – Works Production Assistant

1989–1992 ACE Peterborough – Works Construction Assistant

1992–1993 ACE Peterborough – Permanent Way Renewals Assistant

1993–1995 Assistant Area Civil Engineer (Works) Inter-City Peterborough

1995–2008 Director, Bridgeway Consulting Limited

2008–2014   Civil Engineer, Bridgeway Consulting Limited

2022–2023   The last job. Checking and approving bridge examiner's reports

Photographs

All photos in the book are from my own collection unless otherwise indicated. These include record photos taken whilst working for British Rail

PROLOGUE

I joined the railway industry in 1964, aged 17 and finally left it in 2014 having served for just over 50 years. Since then, I have done some occasional work for Bridgeway Consulting until I finally threw in the towel in May 2023, exactly 59 years after I started work in the rail industry.

This book describes my career and highlights some of the aspects of railway civil engineering that I hope will be interesting and sometimes amusing.

Inevitably it is partially chronological but I have interspersed those chapters dealing with specific time periods with others which describe interests that have lasted for most of my career.

Since I made very few notes of my activities at the time, this book represents my best recollection of events and inevitably, my memory may not be completely accurate, therefore, I beg your forgiveness in advance for any errors. It is not supposed to be historically correct in any case but I will always welcome comments and corrections. There are chapters in the book where I have used the "industrial language" and comments of the worksite. I hope none of this will offend the reader, or, indeed, my workmates.

I hope that my love of the work shows through and I am proud and thankful to be able to say that I enjoyed (almost) all my railway service.

Enjoy!

Brian Maddison C.Eng. M.I.C.E

Chapter 1

Early Days

I was born in the summer of 1946 in Old Fletton, Huntingdonshire. Old Fletton is, in fact, a suburb of Peterborough but located in a different county. This was crucial to the direction that my life took because the education system of Huntingdonshire was quite progressive, and this shaped my future.

Since it was just after the end of WW2, times were fairly hard, so although my parents did not class themselves as poor, there was little money to spare and few luxuries. Dad was a skilled motor mechanic in a steady job, and he spent his spare time looking after the posh cars of some of the local farmers. Unfortunately, payment for this work was often in the form of a couple of rabbits or a hare which did little to help the cash situation and gave my Mum a lot of extra work to do. Strangely enough, the farmers never parted with any of the pheasants that they shot.

We lived in a rented "end of terrace" house that overlooked the church. It had a good length of garden which was mainly used for growing vegetables in the rich black fen soil. At the bottom of the garden, we kept a few chickens from time to time. My sister had been born before the war and there was another child that died at birth. I suppose that my parents did not want to bring up a new baby during the war so by the time it ended, and I came along, there was about 14 years between me and my sister, so I had no young siblings to play with. I am, of course, assuming that my birth was planned and that I was not just an afterthought although my Dad once said that my conception was the result of a Home Guard reunion party. I think he was joking but I have never been quite sure. Whatever, as a small child, I felt loved by Mum, Dad, and big sister. Mother did no paid work in the days that I was young. It was not thought to be "right" for a wife to work and besides, most of her time was taken up with housework and dealing with a kitchen full of dead rabbits and hares. However, she did volunteer at the local library. She was a rather severe lady, and any disciplining was done by her, although she rarely resorted to physical punishment. Dad was a lovely man, very quietly spoken and it was he who found time to play with me and take me on trips sitting on a small saddle attached to the crossbar of his bike.

There were a number of things that happened during my early years that could be said to have begun to form in me the idea of a career as a railway engineer, even at a very young age. Old Fletton is very close to the vast expanse of clay pits and brick works that formed the Peterborough brick industry, although few now remain. It sits on a deep seam of London clay that was excavated locally and, at the other end of the seam, at Stewartby in Bedfordshire. Most of the yards were owned by the London Brick Company and they produced "common" or "Fletton" bricks used in house building. During the time that I was living there, there was a major expansion of the industry to cater for the post-war housing boom. Just after the war, Italian migrants arrived to work in the brickyards. They were initially located in new accommodation blocks and houses, located about 2 miles away from us. The facilities included a swimming pool (outdoor and unheated) which was open to all the children of the area to use for a nominal sum. It was also used as our school pool. It was these brickyards that were my main playground for many years. The nearest yard to us was LB1 which was in full production although the clay pits surrounding it had long been worked out. Instead, the clay was brought several miles to the works in wooden rail tubs towed by a continuous cable system. The quarries around the works, known locally as "knot-holes" had been allowed to flood and formed havens for birds and other wildlife including many species of fish. Health and Safety was not a phrase in common use and there were few restrictions on where the public could go. We played happily around the kilns, crossed over the moving cables towing the clay trucks and fished wherever we liked. Thinking back now, it was a paradise for a child growing up and no doubt contributed to my sense of adventure which I still have to this day.

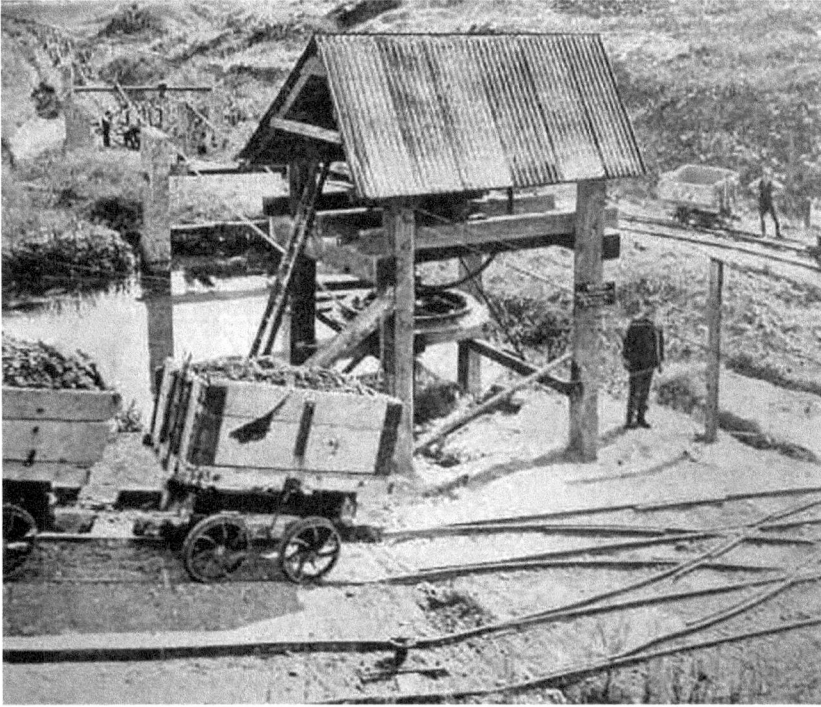

Clay trucks at Hick's brickyard near Fletton. The trucks are moved by a continuous cable.
Photo by kind permission of Peterborough Images Archive.

My first memories of "going down the knot-holes" were long before I was allowed to go there alone. Mum announced that we were going "sticking", put me in the pram and pushed it down to the wooded areas surrounding the knot-holes where I would help collect sticks for the fire or just explore the secret paths through the trees while Mum did the hard work. The sticks and branches were loaded into the pram alongside me to be taken back home. My earliest memory is sitting in the pram surrounded by sticks. I can even recall the smell and feel of the plastic lining of the pram. For me it was fun and an adventure, but I suspect for my Mum it just represented hard and somewhat demeaning work.

One day, when I was still very young – almost certainly a Sunday because he worked until about 1 o'clock on a Saturday – Dad asked me if I would like to go and see the trains. I must have seen trains before on trips on the bus to town, but this was my first recollection of being taken anywhere to watch the trains go by. I was certainly pre-school age, so I was probably about 4 years old, and I proudly walked with my Dad down the lane to LB1 brickworks. There we paused to feel the heat from the kilns and look at the railway trucks waiting to be loaded with bricks. Past the kilns we followed the track alongside the cable wagon-way, which stood silent on a Sunday until we came to the bridge that took the footpath and the wagon-way over the GN main line to London. Today we would call it the East Coast Main Line but that was not a term used in the 1940s.

Once there we climbed through a couple of strands of wire fence and sat on the top edge of the cutting slope to await trains. It was a hot day, and the grass was tinder dry. While we waited for a train to come, Dad lit up his pipe. On a Sunday morning in those days, trains were fairly sparse, even on the main line, so to amuse me, Dad made a wonderful paper airplane which proceeded to fly all the way down to land on the Up goods line. Undaunted, Dad climbed down to recover it while at the same time warning me of the dangers of going on the railway line. A case of "do as I say, not do as I do", one might say. I think on the same day, or maybe on another trip, Dad set fire to the cutting slope grass when he raked out his pipe and we had a fine old time beating out the flames. He seemed to me to encourage mischief and adventure and in my early childhood, he was a real hero to me. Sadly, when I grew older, ill-health robbed us of a lot of happy times that we could have had together.

Much later, my nephew, Paul told me that my Dad (his Granddad) took him on the same adventure and also set light to the grass on that trip as well. I don't remember what trains I saw on that first trip, but I revisited "brickyard bridge" as we called it on many future occasions, initially with Dad, occasionally with my sister and then on my own or with mates. Over those years I saw all the main classes of LNER express locomotives together with a range of mixed traffic and freight locos. My favourite always remained the A4 Pacifics because their appearance was so distinctive. I also saw 9F freight locos, some new from works and later the famous blue "Deltic" diesel. I remember being in class in my junior school learning about river deltas and telling my teacher about the new "Deltic" loco. However, I didn't know the reason for the name was the configuration of the engines and he could see no connection between a railway engine and a river delta, so I got told to sit down and not be so silly!

Once I went to my favourite spot to watch the trains but when I arrived, engineering work was taking place on the Up goods and Up main lines. The ballast cleaner was working, and I climbed down the cutting slope to the trackside to watch it. Track ballast is the bed of stones under the sleepers of the track and over the years it gets ground down and dirty so that it no longer provides proper support. Ballast cleaners are fascinating machines that use a cutter-bar and chain excavator working directly under the machine to remove old, dirty ballast. This then passes through screens on the machine before suitable sized ballast is returned to the track and spoil is loaded by conveyor belt onto adjacent wagons. The whole operation is noisy, dirty, and exciting to watch, it's a bit like digging out the ground that you are standing on without falling into the hole although, occasionally this does actually happen, and the machine comes off the rails. Maybe this excitement was also pointing me towards a future career.

As I grew older and more adventurous, I became a "train spotter" and spent many happy hours collecting train numbers and generally enjoying being on and around the railways. It has to be understood that in the 1950s a very large proportion of school children were doing the same thing, and it was not regarded as the rather "naff" thing to do that it would be today. There is a famous photo that has appeared in several railway books showing hundreds of kids in belted raincoats and school caps at the end of King's Cross, platform 10. Given the opportunity, one of them would have been me. There was also such a wide range of different engines to see from all the pre-nationalisation companies and the newer BR standard types. Most of these were, of course, steam but there were also a small number of prototype diesels that appeared from time to time, plus the famous DC electric locos of the Manchester – Woodhead – Sheffield route. My active interest lasted until I was about 14 but even after this, railways continued to hold a certain attraction. Most train spotting, of course, was done locally and a favourite place of mine was on the waste ground next to where the GN main line crossed the Midland Region lines out of Peterborough East Station. From there, we could see every train in and out of London. We also saw all the trains travelling to and from Peterborough East to Wansford, Northampton and Rugby. The last 2 lines are now closed beyond Wansford tunnel but survive as a part of the preserved Nene Valley Railway. The other place to go locally was New England shed. Visiting the locos on-shed on a Sunday morning wasn't difficult. Even though I never had a "permit" to visit, my Dad's encouragement to mischief came to the fore as I dodged around locos and wagons to avoid anyone in authority. On weekdays, a trip to the North end of the shed was also popular to see an ex-works loco from Doncaster come down on a proving run.

Silver Fox on New England Shed
This photo was taken by me on my Brownie box camera at the age of
about 12 but lay undeveloped for 20 years until my future father-in-law
took it to work (The Leicester Mercury) and developed it.

Another loco shed worth a visit was March and this was usually done by
bike (about 15 miles each way). March loco was much more difficult to gain
access to but there were plenty of vantage points that overlooked the engine
yard. Trips out by train were sometimes undertaken. Child tickets were
cheap and within pocket-money range. So that we could see locos that
normally never visited Peterborough we would travel over the Nene Valley
Line to Wellingborough (for the Midland Main Line) or Rugby for the
Euston Main Line. At Wellingborough we walked from the LNW Station up
to the Midland Station and to Wellingborough shed where there were always
"Crosti boiler" 9Fs on show. Most of the expresses on the main line were
double-headed following the Midland small-engine policy. At Rugby we
were treated not only to the big LMS Pacifics but there was also a vantage
point, well-known to train spotters from where trains on the GC main line
could be seen crossing the LMS main line on the "birdcage" bridge, so called
because of its lightweight steel lattice girders. Most of these trains were the
famous "windcutter" freights but there were also passenger trains to London

Marylebone and the occasional cross-country train with a GWR engine in charge.

On a couple of occasions, I organised coach trips to visit loco sheds around the country. The bus company seemed quite happy to take an order from a 13-year-old boy for the day's hire of a coach and driver. My trips were generally far too ambitious, and we only achieved most of our objectives by being hours late home. Loco works open days were also another great day out. My best trip and one which, fortuitously, led to my travelling over lines now long closed was to the Derby Locomotive works. I caught a train, with several other enthusiasts, from Peterborough to Grantham where we changed and onwards to Derby Friargate, via Nottingham Victoria. When the train pulled into Nottingham London Road High Level, all the others got out because, they said, it was quicker to catch a train from Nottingham Midland to Derby. I didn't realise this and decided not to trust their judgement but to stay on a train that was already heading in the right direction. It was a slow journey, and it was also a long walk from Derby Friargate Station to the Loco Works so at the time, I thought I had made the wrong choice. However, much later, when I came back to Nottingham to work and the line from Nottingham to Derby via Ilkeston had been long closed, I realised how lucky I had been to have travelled a route that crossed the Bennerley Viaduct, which still stands and is now a "listed" structure and has a new footpath and cycle route across it.

I joined the Army Cadets when I was about 14 and when this became my main passion, trainspotting faded out of the picture but as a final fling, I spent 2 days in South Wales, travelling by train and sleeping in station waiting rooms. On this trip I traversed the "Heads of the Valley" line across the Crumlin viaduct before the line was closed. I attended the local "Board" school as an infant and junior until I was 10 years old. This school was located next to the GN main line so journeys to and from school were often delayed if Fletton Junction signals were "pulled off" as we crossed the bridge over the line. I managed to pass my 11-plus exam and went on to Old Fletton Grammar school which was one of only 3 such schools in Huntingdonshire. It was co-educational and from the outset it had been a county secondary school so there were no associations with former fee-paying establishments and no suggestion of any class hierarchy amongst the pupils. Each village in the catchment area of the school was allocated a number of places and I am told that Old Fletton had 4 allocated places, so I was very lucky to be one of the chosen few. I cannot say that school was the happiest time of my life. I always seemed to be struggling to keep up with the rest of the class.

Nevertheless, I was reasonably successful and although I didn't carry on to the sixth form, I did leave school a little before my 16th birthday with 7 GCE 'O' levels. I can see that the selection for grammar school at 11 years old is a brutal process and one which in many cases determined the whole direction of a person's life from a very young age and in principle I believe that the comprehensive school system is better. However, the grammar school system did result in many more children from disadvantaged backgrounds going on to achieve great things than appears to be the case today.

From the age of 14 till 18, I spent most of my spare time with the Army Cadets so my interest in trains generally waned but there were a few memorable trips during that period. An army cadet summer camp in Warcop saw us all travelling by steam-hauled troop train from Kettering through Leicester and Leeds and on to the Settle and Carlisle line over Ais Gill. Another camp was held in Germany, and we travelled by train and ship via Harwich to Osnabruck. I have some photos of the cadets on a visit to Osnabruck marshalling yard, but the highlight of the camp was the chance to climb inside an army tank. There was also a school trip to Eastern France, again by train with a return journey from Nancy to Paris behind a big pacific loco.

With the Army cadets in Germany

I left school without a clear idea of what career I should be aiming for. I took temporary employment with the local Health Executive and was offered a permanent job there, but I was determined that whatever I did, it would involve more than office work. I then got an offer of what I thought would be my ideal job. The local firm of Mitchell Engineering were involved in the construction of Windscale Nuclear Power facility at Sellafield, and I was offered a job there as a trainee engineer. I must admit to having mixed feelings about leaving home at the age of 16, but I was ready to give it a go. However, it was not to be, the firm must have had similar reservations, and I received a letter cancelling the job offer because a senior member of the company had said I was too young to go.

Why I didn't consider applying to the railway at this time, I am not sure, but after several rejections from construction companies, a family friend who ran a small local building company offered me a job in the office and the chance to study building construction on day-release at the local technical college. Unfortunately, the job was incredibly boring, mainly involving processing invoices and answering the telephone. None of what I was doing at work helped me with my college studies, so I fell behind the rest of the class and eventually started skipping lessons. It wasn't long before this was brought to the attention of my boss who, because he was a friend of my parents, simply suggested that I find another job rather that sacking me on the spot as he would have been entitled to do.

This was a pretty low point in my life. I felt that I had let a good number of people down, including my parents. Both Mum and Dad were unwell, mainly because of Dad's problems and I am sure that I was adding to them. Fortunately, while discussing this all with a friend who worked as a clerk on the railway, he suggested that I try to get a job with their engineering department, so I wrote a letter to the District Engineer, Peterborough and managed to get an interview. Actually, getting the job was not that easy but my time spent in the Army Cadets helped. Like many senior railway engineers of the time, the District Engineer, Ted Labrum, had been a Major in the Royal Engineers and my time spent in the Army Cadets and my achievement of reaching the rank of Cadet Sergeant and a getting a Duke of Edinburgh's Award clearly impressed him. At the same time, he was very much less than impressed with my work record since leaving school and if took 3 interviews before he agreed to take me into the drawing office. Even then it was for a probationary period of 6 months during which time I was classed as a "temporary tracer". It was also agreed that my time spent at college studying Building Construction would be written off and I would start again on an ONC course in Civil Engineering. So began a career as a Railway Engineer that lasted over 50 years and I am pleased to say that in all that time there were only two short periods, each of about 6 months when I was not thoroughly happy and fulfilled in the job that I was doing.

Chapter 2

Peterborough – Part 1

I joined the staff of the District Engineer Peterborough in May 1964 at the age of 17 and embarked on a career that would last a lifetime. The engineer's yard was situated north of the station on Mayors Walk. The main offices still exist along the road side and are used as a traincrew signing-on point. The remainder of the yard is now a station car park and most of the buildings have been demolished. As well the drawing office, the yard contained all the functional buildings one would expect, including a two-storey carpenter's shop and workshops for other trades. There was also an internal railway which was shunted by a small "Sentinel" steam loco. My designation as a "Temporary Tracer" was a device used so that if my work was unsatisfactory in the first 6 months, I could be sacked without a formal disciplinary procedure. I was on tenterhooks as the end of my probationary period approached and indeed (because of a forthcoming re-organisation) it was actually several months after the 6-month period that my permanent employment was confirmed.

The duties of the post were to undertake whatever design, drawing, surveying and technical supervision work was required. The office was quite small and was broadly divided between "Works" (the structures) and "Way" (the track) under the control of the head of the drawing office, Arthur Rodgers, who later went on to run the Civil Engineering Training Centre at Watford for many years. A desk was allocated to me next to the redoubtable Betty Saunders who was a good engineer and who rather took me under her wing and was probably the main reason why, later on, I rejected the "sexist" attitude of many Civil Engineers. Our part of the office was next to the plan room which was attended by Bill Clark. Bill had lost his leg in a railway accident at Grantham. I had only been at work a couple of days when he came to me and said, "Make sure you join the Union, Brian, because *when* you lose your leg, they will look after you." I have emphasised the word *when* because it sounded almost inevitable that some major injury would befall me sometime in my railway career and, indeed, the safety record of the railway was pretty poor at the time, so it was probably fairly close to the truth. I was loosely a part of the building section at first but did jobs for all of the "works" side of the office as well as being inducted into basic P-Way duties including de-stressing and setting out track lifts ahead of the tampers. The latter were essential skills because on any given weekend there was always a need for technical staff on relaying sites.

The staff of Peterborough District Engineer's office taken in about 1963, just before I joined. Betty Saunders is on the far-left side of the picture on the front row.

The first job that I was given to do was counting the bricks in a number of bridges on the closed Holwell Branch. Fortunately, I was able to do this from the original bridge drawings. Apparently, it was required to provide a valuation of assets that were declared redundant. For the whole of my time with British Rail, the plan room could be relied upon to provide details of all our bridges in the form of original "as built" drawings and drawings of any alterations and repairs that had been subsequently carried out. Many of these were lost by an appalling act of official "vandalism" during the privatisation process when they were just thrown into a skip and lost. Others were taken to historical archives and were equally lost to working engineers. Whenever other duties allowed, I was expected to practise my drawing skills using pen and ink on blue drawing linen which required chalking before putting pen to paper. This work would then be inspected by Arthur Rogers who was a hard man to satisfy. This practice stood me in good stead throughout my career and as I got better, I began to love that aspect of the work and I found that I could lose myself for days on the production of a detailed bridge drawing.

During the 2 years or so between leaving school and starting with BR, the work that I had done was very basic and always under direct supervision. What a difference it was when I joined BR where I was expected to hit the ground running and immediately began to be given jobs requiring make my own decisions. A good example of this was a trip to Boston Docks. During the sixth week of my employment, I was called into Arthur Rodger's office late one afternoon and was told that I was to go to Boston by train for a meeting to discuss the weighbridge. I set off the next day, looking forward to a pleasant trip and a look round the docks and being just a little apprehensive because I had discovered that I was the only person attending from BR and that the file I was given showed that the Docks Engineer had concerns about the track over the weighbridge as well as the building itself.

When I got to Boston, I elected to walk down the track to the docks, over the rail swing bridge so as to get a look at the weighbridge on my way to the offices of the port manager. There I was shown into a large conference room to be introduced to several people including the Port Manager and the Port Engineer as "Mr Maddison, the representative of British Railways". I have no idea what they thought of me, these 2 senior managers looking at a 17-year-old "boy". Furthermore, it transpired that the Port of Boston, which was owned by the local authority, had a significant and long-standing grudge against British Railways. King's Lynn Docks was Boston's rival port on The Wash, and this was a nationalised port with close associations with British Railways through the British Transport Commission. We stood accused of favouring King's Lynn over Boston in the provision of empty wagons for loading, freight services, stealing their teacups, eating their fish, ravishing their women and all manner of other things. I struggled with all this, but I was helped out by the Port Manager who steered the meeting back to weighbridge and graciously accepted my assurances to take back full details of the problems to my boss where they would be fully dealt with. The swing bridge remains in operation today with a healthy number of trainloads of imported steel being railed from the docks to the West Midlands. In this respect, Boston has triumphed over King's Lynn because their Port is no longer connected to the rail network.

After a few more basic jobs, the small bridge section in the office, which consisted of Tim Green and Derek Smith, decided they needed the assistance of a junior and got me an "unofficial" transfer into their section. That is when my love of the work really took off. I had not been there more than a few weeks when I was told to accompany Tim on the reconstruction of a multi-span underbridge, which carried the East Coast Main Line over a watercourse to the south of Tallington level crossing. Tim's idea of making sure I was properly trained was to get me to site and then let me loose with engineer's level and tape measure to control the positioning of all new parts of the bridge whilst also making sure he had a supply of "Senior Service" cigarettes. It was a daunting task, made no easier by an irascible foreman bricklayer. When using a level to set the beams, one must read the number upside down and make sure you are passing on the correct instructions to either raise or lower the mortar bed for the beams. It is very easy to make a mistake and the bricklayer had added 50mm of extra mortar to the bed before I realised I had it wrong and he had to take it all off again. Words like "young whippersnapper" and worse echoed in my ears and I never made the same mistake again.

Though I was learning quickly through my experiences on the railway, I was still a young man with a lot to learn, including dealing with bad weather on the road. I was riding a motorbike at the time that this bridge was underway and on the second weekend I set off from my house in thick fog. There was no improvement to visibility on the 10-mile journey, though, strangely, other traffic seemed to be making light of conditions and passed me at some speed. I cursed their foolhardiness. When I arrived at site, I opened my visor to reveal a perfectly clear sky. The fog had all gone but had left a misted-up visor! This was also the first time that I came across the wonderful steam breakdown cranes which we used for engineering work and my love affair with them lasted for 25 years but these magnificent machines will be discussed in later chapters. The concrete beams for the Tallington bridge came to site by rail from Boston via New England yard.

On the first weekend, we had started work on the bridge when a full train of beams passed the site at speed heading north on the Down Main behind a Stanier 4MT loco. We then realised these were our beams and the train had to be stopped at Grantham and returned to site an hour or so later. This was my first bridge reconstruction and the only one I ever did "under supervision". After that, I was on my own to face the problems of having a big hole where there should be a bridge and only 6 hours until the first train was due! The hands-off approach of Tim Green to let me get on with the job while he watched from the sidelines inspired my confidence when it came to being responsible myself.

Not long after I started, I was being recruited to work weekends. You would think that an 18-year-old would have much better things to do on a Saturday night than go to work, but the incentives were great and I soon fell into the rhythm of railway civil engineers which was to have a good night out on a Friday and a lie-in on Saturday morning to prepare for a long shift at work on Saturday night through to Sunday. At first, for me, I was not driven by the extra pay because by moving from a £5.00 per week job to the £10.00 per week that I got on the railway, I had already doubled my wages. However, for married men – including me later on in life – the weekend shift, paid at "Time and Three-quarter", represented the difference between a relatively good lifestyle and genuine poverty. The real incentive for me was that all the major works happened at the weekend so I lapped up every shift I could get with the bridge section and whatever else came my way, including numerous rail stressing shifts. When rails are installed in the modern railway, they are welded together to form a continuous ribbon of steel from end to end of the line. Steel expands in hot weather and contracts in cold. Too much expansion and the weight of the track and ballast cannot contain the compression forces and the line will buckle. Too much contraction and the rails will break. To prevent these problems, rails are installed at an equivalent temperature that is somewhere between lowest and highest expected temperatures throughout the year.

In the 1960s, the way that this was achieved was to physically heat the rails using propane burners pushed along on a trolley. The same effect is now achieved by stretching the rails with hydraulic jacks. In both cases, the engineer's role on site is to ensure that the rails are welded up after the correct extension has been achieved throughout the length of the rails. It is a fairly simple job, but things go wrong, including the propane cylinders falling off the trolley and on a cold night on a falling thermometer it can be impossible to get the rails to expand because the air temperature is falling too quickly. While doing the de-stressing shifts, I would also make myself useful around the whole relaying site and so I learned the whole relaying process. It was also on weekend shifts that I learned that the working environment could be a very harsh place, particularly for the contractor's labourers that we often used to supplement the full-time railway staff.

I had a bridge job at Grantham that ran for several weekends and involved removal of rails, sleepers, and ballast over the bridge and installing a waterproof layer that it was hoped would reduce the amount of water seeping down onto the road below. Much of the work was carried out by hand with gangs of "Tersons" men unloading the new sand and ballast before the tracks were replaced. I had become friendly with one of the men – he was Irish like most of the gang – but one night while I was watching him, he didn't seem to be working with his usual gusto. I mentioned this to the ganger in case there was something wrong with him, and the immediate response was that if I wasn't satisfied that he was pulling his weight, then he would be sacked immediately, and the ganger set off to do just that. I had to run after him to stop him and explain that his work was fine, and he was just not as cheerful as usual.

Later, I mentioned this to a colleague who told me this story, which he swore was absolutely true. Apparently, the same gang was travelling to site in their bus when a fight broke out. The ganger broke up the fight and identified the man who he thought was the culprit and told him to get off the bus. The bus was way out in the countryside, it was late at night, and it was pouring with rain, but the man was sacked and told to make his own way home. More than that, before he left, the ganger said to him. "Whose wellington boots are you wearing?"

"Tersons," came the reply.

The ganger said, "Well, you can take them off," and they left him there in his stocking feet – harsh justice!

Years later I was travelling on a train passing Ratcliffe-on-Soar power station and I was talking to a man who sat next to me. He told me that he was brought over from Ireland to work on the construction of the power station in the 1960s. Along with all the other men from Ireland, he was housed and fed on site by the contractor. They were all told that the site was too far from a town for them to be able to go anywhere for recreation and they remained there for the length of their employment, which in some cases could be 5 or more years. It was only later when the work had finished that he realised that they were less than 10 miles from Nottingham and Derby and within walking distance of a village with a pub. This was the 1960s and men were still being treated like slaves!

A more comfortable weekend shift was structure gauging and while I was at Peterborough, we undertook the gauging of the structures along the GN and

GE joint line from Spalding to Sleaford. Clearly, the railway must take steps to ensure that when a train runs along the line, it doesn't hit the bridges and other lineside structures. This is achieved by ensuring the structures along the track are positioned sufficiently far away from the line and that the rail vehicles are not too big. Today, this is a sophisticated process involving fixed survey grids, GPS and computer programmes which have enabled the tolerances to be much reduced and therefore bigger loads can be carried and passenger trains can be built that do not require a big step between train and platform edge. In the 1960s, our main piece of equipment was the "gauging van" which was, in fact, a 4-wheel "Grampus" wagon fitted with a fixed wooden hut that contained a pointer on an extending and rotating arm that would be pushed out to touch all the key points on a lineside structure. Additionally, low structures were measured from the ground and the cant of the track and curvature of the line were also measured. The information, which was produced as a series of overlying drawings showed the structure gauge of the line and was used to check that no structures impinged on the safety envelope between structure and a normal train. They were also used to calculate whether large "out of gauge" loads could be carried. The team to do the gauging normally consisted of two staff from the drawing office plus Bill and Bob, the two office chainman, who were needed to assist with the measurement of curvature of the line. The wagon was pulled by a loco and there was a break van with guard on the back.

On the first week, we met our very short train in the sidings somewhere in Peterborough. New England had generously supplied us with a brand-new Brush type 4 loco (now called a class 47). We travelled very quickly to the first site but then the problems started. At bridges and tunnels, the gauging could be quickly completed but stopping this very large engine in the exact spot so that the pointer could be pushed out to touch an isolated signal post proved very difficult and time consuming, so progress was not as good as anticipated. Nevertheless, some work was completed but for next weekend, we requested a smaller engine. New England again responded to our request with alacrity by providing a Class 08 shunter which could position us accurately with ease but had a top speed of about 20 MPH and took an age to get to site!

The railway is very rural and one Sunday morning we were on the Down line between Gosberton and Blotoft when I looked back to see the unmistakable sight of a plume of smoke from a steam engine rapidly approaching us from the rear on *OUR LINE*. Not only did we have a possession to do the work, but the line was closed to traffic anyway on

Sundays so a degree of panic ensued. By this date there was little steam left operating from Peterborough so I had a momentary thought that it must be some sort of ghost engine from long ago and it would just pass straight through us, then a second more plausible thought that an enthusiast's special marking the end of steam might be on the line. There were quite a few of them about at the time. The guard was already making his way back towards the approaching train with his red flag. Fortunately, it turned out to be the relief engine crew for our own train who had been sent by light engine instead of the usual road vehicle.

I have to say that the work was a very pleasant job in good weather with occasional breaks for cups of tea and blackberry picking but on one occasion our guard came away with a bigger prize. We were gauging in a cutting when we looked down from our wagon as a rabbit raced by along the cess, closely followed by our guard. He was followed by ironic cheers from the rest of us, never thinking that the rabbit would be caught but it made the mistake of diving into a hole in the cutting slope which turned out to be a discarded drainage pipe with no opening at the far end. The guard walked triumphantly back towards us carrying a fine rabbit for the pot.

Just east of Peterborough on the line to March, there is a bridge called "Black Bridge" over the old course of the River Nene. It used to be a large, multi-span bridge with 9 tracks, constructed entirely of timber. The bridge was about 15 minutes' walk across the fields from my house and when I was younger it was a favourite spot for fishing and generally messing about by the river. About a year before I joined the railway, a huge plume of smoke and flames erupted from the bridge and all the local youngsters, including myself, ran over to watch the bridge burn spectacularly until only the stubs of the timber piles remainder protruding above the water line. The task of re-building the bridge fell to the Peterborough Engineers office which I joined later. By the time I started work, the bridge had been rebuilt in timber and had re-opened to traffic, but I was involved in sway and deflection testing of the structure which required the reading gauges set up to monitor movement as trains crossed the bridge. The story of the re-building of the bridge only served to further enthuse me to the career that I was embarking upon and is an illustration of the skill and teamwork and resources that could be called upon from the railway "family" but also the power that could be wielded on the railway's behalf. For instance, within a few hours of the fire being extinguished, it had been decided that quickest way that the bridge could be re-built was in timber on a more-or-less like for like basis albeit with only 2 tracks. Obtaining sufficient heavy timber was clearly a priority and when it was discovered that a ship loaded with the right sort of timber was crossing the North Sea, the whole load was requisitioned, and the ship was diverted to Boston Docks for unloading onto rail wagons. There were many photos of the progression of the works up in the office and also pictures of members of the staff receiving their official toy duck awards for falling in the river! It was generally acknowledged that the work was well done and in double-quick time and I sought to emulate this performance throughout my career.

Away from my work duties at Peterborough there were many other interests to attract my attention. I joined the Permanent Way Institute and attended meetings. This was useful not only to gain knowledge but also to further a relationship with colleagues from around the district, especially those from outposts such as Boston, Grantham and Colwick (near Nottingham). There were also perks to be had such as additional free rail passes that were granted for attending First Aid classes and things called "firewood" orders. I had already done some first aid training with the Army Cadets, so I was pleased to go to the classes. I continued to be a "first-aider" throughout my time on the railway and occasionally dealt with minor injuries, but it was in the first few weeks of starting the First-Aid course that I had my one and only serious incident.

I was walking, with the office chainmen across a field to get to the side of the line between Orton and Wansford to carry out survey work when we were hailed by a man who was with his small son. We found that it was a party of 3, father, son, and grandfather who were heading towards the river to go fishing. The old man had collapsed and appeared to be having a heart attack. I was the only first-aider amongst us so while I administered CPR, one of the chainmen ran to the signal box to summon assistance. It was about 20 minutes before an ambulance could find us. Frustratingly, we could see it driving up and down the main road trying to find the field entrance. I did my best with mouth-to-mouth and chest compressions which I kept going until the ambulance man arrived. I think I really knew that the man was dead but from time to time, probably due to the air I was pushing in to him, his chest would heave. Because of this and my own lack of experience, I didn't dare stop until I was relieved. The ambulance man then said – and I quote verbatim – "I don't know why you called us; he is already dead." I felt somewhat useless at the time but at least I tried, and I did get a letter of commendation from the St John's Ambulance service.

Stealing from the railway by staff was considered a serious offence and all cases were vigorously prosecuted by the British Transport Police. In fact, most railwaymen felt that the railway police force was keener on catching railway employees doing something wrong than they were in pursuing genuine criminals. This was made very clear to all new starters when they joined the company and was also emphasised in the "Rule Book". It was well known that people had been sacked just for picking up pieces of coal that had fallen from engine tenders.

To ensure that there was a route for railwaymen to acquire surplus materials legally and not be accused of stealing, there was the ubiquitous "Firewood Order". Staff could apply to be issued with an order which would enable them to purchase firewood and other items for a nominal sum. It was generally recognised that 10 cwt of firewood at "One and tuppence" a cwt was about the minimum price to be paid. I got a very useable shed which I erected in my parent's garden for this princely sum. It was a prefabricated shed in the yard at Mayors Walk that was going to be thrown away but which I reckoned could be dismantled and re-used. Once the form was issued and a supervisor signed it to say that the material was available, it had to be taken to the nearest booking office where payment was made, and the form stamped accordingly. I was also able to agree transport home on the railway lorry providing I arranged all the dismantling and loading myself. The shed proved excellent for storage of my motor bike, amongst other things. There were tales of all sorts of expensive items being bought on a firewood order, some of which were certainly questionable, but I think a redundant shed was a fair use.

While I was at Peterborough, I was made aware of circular offering places on a month-long Outward Bound course to young railwaymen from the Eastern Region of BR. There were apparently 4 places available, 2 at sea schools and 2 at mountain schools. I duly applied and had to write an application stating why I wished to go on the course. I was eventually short-listed and attended a final selection at Liverpool Street HQ where I had to write an essay. I was successful, possibly due to my choice of my Army Cadet experiences as a topic. I was allocated a place in early June at the sea school in Burghead, on the Moray Firth in Scotland.

The journey north was impressive. I travelled overnight from Peterborough to Edinburgh then caught a morning train to Aberdeen. These were still steam-hauled at the time and the train was equipped with a restaurant car serving a full breakfast including porridge, which I was eating on my first ever crossing of the Forth Bridge. Once in Aberdeen we were accommodated on a local train to Elgin which ran via a coastal route that has long-since closed. I have to say that Outward Bound courses are really tough. For the first week, every exercise was done in boots, including cross-country runs! After that week, however, when we reverted to plimsolls, I felt as if I could run like the wind and jump over a church spire! Every morning at 6 o'clock we ran about a mile to the sea, which was very cold, jumped in for a quick swim and then run back for breakfast. The school had a small 3-masted schooner called the "Prince Louis", which was used for the main expedition at the end of 3 weeks training. My group started the expedition by rowing and sailing small cutters along the coast towards Inverness and from there we proceeded on foot, crossing to the west coast via a number of peaks. Two nights were spent in the mountains and on the second night we endured torrential rain which left our tents cut-off and surrounded by water. We had to wait till around midday before it had subsided sufficiently for us to carry on, but we still had to get the allocated distance covered so the afternoon and evening were spent on a forced march to the west coast where we met up with the "Prince Louis".

There were only a couple of professionals in the crew (Captain and Boatswain) plus a rather grumpy cook who was on secondment from the Royal Navy. In addition to all the general duties, I was allocated the title of "Captain of the Heads", which I was initially quite proud about until someone explained to me that I was the one who cleaned the latrines. Heading north towards Cape Wrath, there was no wind, so we had to use the motor. Unfortunately, the round-bottomed ship rolled badly without the stability of sails so most of us felt quite ill, but we were worked hard the whole way and just had to suffer. As soon as the wind freshened, the sails were hoisted, and seasickness disappeared. The wind continued until it developed into a gale with lashing rain as we passed through the Pentland Firth. I was on lookout in the bows when directly in front of us loomed a large freighter. I shouted a warning to the bridge, and we took avoiding action. I thought that I had done OK, but I was given a reprimand for not using the correct wording in my message. To be honest, I was so frightened that I am surprised that I managed to make any sound at all.

As the storm strengthened, we made for shelter in Scapa Flow where we saw many of the wrecks of the scuttled German Fleet. We also passed close to where HMS Royal Oak had been sunk by a German U-boat in 1939, killing most of the crew. The boatswain of the Prince Louis was serving on the Royal Oak when it went down. He showed us the oil that was still seeping to the surface from the wreck and told us he only survived because he happened to be on deck having a cigarette at the time. We were now late and so when the storm abated, we sailed south through the night along the east coast back to Burghead. That was a magical journey; under light winds and clear skies we spotted dolphins that followed us for miles. I do really think that my month at the school instilled a "can-do" attitude into me that I never lost throughout my career, and it stood me in good stead when working under the pressure of tight possession times.

My Outward Bound group at Burghead.
I am the tubby one in the back row.

The Prince Louis.
We sailed through the Pentland Firth in a storm

The rest of my time at Peterborough was spent working on a range of generally minor building and bridge repair schemes and learning as fast as I could about working on the railway. I also worked for the P-Way department at weekends and I was going to Technical College to gain an Ordinary National Certificate in Civil Engineering. Working life was full of excitement, particularly when on site and being junior in the office didn't stop me from getting some really interesting jobs. One of those was to go as "scribe" for the bridge examiners on the GUZUNDER. You will probably be aware that a GUZUNDER is a name for a chamber pot because it *GUZUNDER* the bed! However, this GUZUNDER was a large inspection platform mounted on a rail wagon. It consisted of a working platform, big enough for 4 people, mounted on an articulated hydraulic arm that could be swung out and down, giving access to the underneath of the bridge that the machine was standing on – invaluable for the inspection and repair of high viaducts and bridges over water. Because it occupied the track, it normally had to be used at nights or weekends but on the lightly used freight line from Peterborough East to Oundle we were able to do the work on normal weekdays.

There was one long viaduct to be examined and some bridges over rivers. I was excited as we set off from Peterborough in the mess and sleeping coach that always accompanied the train and our first stop was at Wansford where the line crossed the River Nene. I was joined on the platform by an operator and a bridge examiner. The bridge was so low over the water that the articulated joint in the arm dipped into the river as the platform was positioned but I thought little of it. When the arm lifted water was scooped up and rushed through the tube onto the platform. I was soaked from head to foot, and it was only then that I noticed that my 2 colleagues had jumped out of the way. I guess that this was a sort of initiation to bridge examining! Later in my career, I worked nights on the GUZUNDER while examining the 80-span Harringworth Viaduct. In the dark, we merrily stepped into and out of the platform on to the top of the parapet (which saved the time of swinging it in and out for every span). Only when it began to get light in the morning and I saw the ground way below me did I develop any fear of the height!

The happy time at Peterborough came to an end in 1965 when Peterborough District succumbed to a re-organisation and I was transferred to King's Cross Division, along with several my colleagues. I reflected at the time that not much more than a year before I had been walking about 100 yards from home to the office to do a thoroughly miserable job for a pittance in wages and that now I would be travelling almost 80 miles to work and back every day to carry out a hugely interesting and responsible job for a decent wage. What a change I had had!

Chapter 3

Woolsthorpe Branch – Bridge 2

There had been a small argument going on in the Peterborough drawing office about which section I should be working on. The building work I was doing for Ivan Haywood and the outside party work for Betty Saunders was pretty routine. The bridge section was busy, and Tim Green was anxious to secure my services. He was also very encouraging to younger staff and keen for them to be given challenging work. Tim was always going to win the day; he had a forceful personality and was destined for higher things. I would say that he probably had the most influence on my early career and I was pleased when our paths crossed again at Nottingham in the 1980s. As soon as he got his way, Tim put me on the project to reconstruct Bridge No. 2 on the Woolsthorpe Branch.

The Woolsthorpe Branch was built in the 1880s to carry Iron Ore from the mines around Denton in Leicestershire to Belvoir, a junction on the Nottingham to Grantham Line. The line as far as Woolsthorpe Sidings was the original branch and a later section from there to Harston was known as the Woolsthorpe Branch Extension. There was also a short branch to Harlaxton Quarries, built in the 1940s. The line was always known by these names on the Eastern Region, but it is also sometimes referred to as the Denton Branch which I think may have been a miss-naming by those Midland Region scoundrels. The line never carried regular passengers but for most of its life, it was busy with ironstone traffic which was transferred from the Ironstone Company's private railways and tramways. Much of the output went to Stanton Ironworks in Derbyshire and other ironworks further north at Staveley. From time to time, loads would travel further to the mighty furnaces at Scunthorpe.

In 1964, the line was busy and mainly steam worked with locos from Colwick MPD. I believe that there was a period in the late 1960s when the line closed for a while but it quickly re-opened and continued into the mid-1970s before production of iron ore ceased in the area. By that time, the traffic was being worked by pairs of Class 20 English Electric diesels from Toton Depot. Bridge 2 carried the single the line over the River Devon just behind the church in the village of Muston. It was a timber trestle structure, the condition of which had become so poor that complete reconstruction was necessary.

The initial site visit was a first for me in many ways. Firstly, it was my first trip with George Flint. There was no personal transport available for drawing office staff and in fact, other than gang buses and lorries, the only transport available in the yard was the engineer's car with his personal driver and an old Bedford minibus which was driven by the bad-tempered George who did exactly as he liked. The former might occasionally be made available to other senior staff but never to the likes of us in the drawing office, so we were left with George and his minibus. There was, of course, a booking system; however, the time that you booked would not necessarily be the time that George decided he would be prepared to drive you. Furthermore, George point-blank refused to drive on the A1, so a journey from the office to Muston, which should have taken less then and hour via the A1 took almost twice as long via Bourne and Grantham town centre.

So, Tim, Derek, and I set off from the office late and by the time we got to site and met with the local works supervisor we were even later. This was my first real important site meeting where I was actually taking part in the planning of the works. I was also responsible for taking notes of what was agreed. Despite being late we got a good look at the site and the methodology for the works before retiring to the Red Lion in Bottesford for my next "first", a Chicken and Mushroom pie. I was probably just 18 at the time and I had already been drinking in pubs for 2 or 3 years but only in rather seedy locals with my own mates in the evening. I don't think I had ever before eaten any hot food in a pub and to me it was a wonderful treat. After pie and a couple of pints we went back to site but then had to set off back to Peterborough when George informed us it was time for him to go.

The plan was to remove all the old timber trestle and construct a new tunnel formed corrugated metal sheeting known as an "Armco" multi-plate arch. The River Devon would flow through this structure and an embankment would be constructed over and around it to support the railway. It was normal practice for bridge re-constructions to be designed at the offices of the Chief Civil Engineer and the drawings then passed to the local district to carry out the works. I think that might have been the case on this occasion although it is possible for a relatively small bridge for the design to have been produced locally. Whoever had designed it, we now had the drawings and we just needed to build the new structure. This had to be done over a weekend from Saturday lunchtime to Sunday night to avoid any disruption to traffic. I was given the role of planning all the possession work, including a method of working, a timetable for the works, the provision of resources and a drawing showing the site layout on which the positions of all plant and materials stockpiles were shown.

The site was quite cramped and there was no road access. Major materials had to be brought in by train and a dragline excavator was tracked in over farmland. Before the weekend, fill for the embankment was brought in by train and stockpiled alongside the line. This will probably have been "spent" ballast brought direct to the site from ballast cleaning works elsewhere. A multi-plate arch bridge is essentially a corrugated steel pipe with the corrugations running around the diameter to provide strength. However, because of their size, they cannot be transported as whole pipes but are delivered in pre-curved pieces to be bolted together on site. In themselves, they are not particularly strong but when surrounded with a properly compacted fill, the compression forces surrounding the pipe form it into a very strong structure. Normally the pipes are built from end-to-end so that any adjustments to the position of the pipe can be made as work progresses but in order to get as much work done before possession of the line was taken, the invert of the arch was constructed first and then each end of the pipe built up to the old timber bridge. The centre section was left to be completed after the timber bridge was removed.

All the planning was completed and after a couple more visits to the site for setting out and checking on the pipe installation with, of course, chicken and mushroom pies at the Red Lion, the "day of the race" approached and the question of what shifts we would work arose. George Flint had to be involved in the planning because it was difficult to get to site without him. I think it was decided that Tim would do the first 12 hour shift and Derek the second. I was to go with Tim on the first shift and stay on site, helping and learning for as long as it was deemed necessary, and I presumed I would probably go back to Peterborough with Tim. We arrived on site to find the ballast train just arriving, being propelled in from Belvoir Junction by a 350hp diesel shunter. This had been specially chosen because when not required, it could be shut down on site whereas a steam loco would have had to return to Colwick for coal and water.

Work proceeded well; tracks were removed in panels and timbers from the bridge loaded into wagons for disposal. By the evening, we were ahead of schedule and ready for a break. Someone pointed out that the Muston Gap pub was just down the line towards Belvoir Junction. We jumped into the brake van and the train drew us down to the back of the pub from where we clambered down the embankment and into the back door. The local clientele were mainly gentleman farmers and the like. They didn't take kindly to a bunch of dirty railwaymen disturbing their Saturday night drinking but apart from a few comments under their breath about nationalised industries, they didn't dare confront us directly. I think we got a feel of what is must have been like when the Navvies came to build the railways but fortunately without any violence. After 2 or 3 pints we left them to their tut-tutting and climbed back on the train to go to work again.

I busied myself doing whatever was needed. Whilst the steel gang took a break, I tried my hand at using the oxyacetylene cutting gear on the bolts of the old bridge. Much later in the shift, it was found that the track panels were not particularly well loaded and if left as they were, there was a fair chance that they would hit a signal post or similar lineside structure on the way back to the yard. The crane was no longer available, so the only choice was to cut off the ends of any sleepers that were sticking out. Having planned the job, of course, I knew where all the equipment was and as I was largely a supernumerary at this point, I volunteered to do the work, even though I never handled a chainsaw before. Now if you have hired a chainsaw from a local plant company, you will know that the only ones available to you have very short blades and that you will be given special safety trousers, hard hat and a visor to protect you from accidents. It goes without saying that we had no safety equipment, which would have been unheard of in 1964. Added to that, the chainsaws we had were driven by an air motor supplied with high-pressure air from a compressor and blade that was at least 4 feet long. It took all my strength to wield it to cut off the sleeper ends above my head, but I loved the work and persevered, without accident, until all the sleepers were cut down to size.

The construction of the new bridge had slowed at this point. There was a lot of tolerance in the bolt holes of the panels, and this allowed the structure to go slightly off line as it was being built, meaning that as the two sides of the bridge approached each other; it became increasingly difficult to line up the bolt holes. Fill was being placed over the bridge as fast as each panel was fixed in place, so it was not possible to move the panels once they were in place. The only solution was to burn a few new holes which is not really to be recommended but did not greatly affect the strength or durability of the structure. The bridge is still there now some 60 years later, and it is possible to walk over it using an unofficial "footpath".

Tim Green was relieved by Derek Smith early Sunday morning, but it was suggested that I stay on for a while to assist and provide continuity. I was more than happy to do this but as the day wore on and I was still there, I did begin to wonder how I would get home. Fortunately, early Sunday afternoon, there was a vehicle going back to Peterborough and I managed to get a lift. That was the end of the longest shift I have ever worked. Including travel, it was over 24 hours which was a long time, even for those free and easy days. I was still living at home and my mother was on the point of contacting the railway authorities to see if I had been killed. I have to say that I enjoyed the whole experience and I wouldn't have missed a minute of the whole thing.

Bridge 2 as it is today. 60 years after construction

Chapter 4

Down to "The Smoke"

For most of my time on British Railways there was the threat of a reorganisation hanging over the workforce. I suppose that this was inevitable while the railways were contracting and under ever-increasing pressure to save money. It is strange that if British Rail didn't break-even financially each year, it was vilified as being a loss-making nationalised industry whereas the huge losses made by today's train companies are protected by government subsidies which allow them to pretend that they make a profit. The first of the re-organisations to affect me was, in fact, being discussed before I even started work and it wasn't long before proposals became official, and we knew that our time at Peterborough was limited. This particular proposal was really just to reduce the number of Districts to compensate for the fact that many lines had been closed and fewer office-based staff were required. However, as usual, it was dressed up as a major improvement to the way we worked and to emphasise that, we would cease to be a "District" but become a "Division". Subsequent re-organisations proposed name changes to "Territory" and "Area" amongst others. There is a famous quote by Petronius Arbiter, a Roman Courtier during the reign of Nero which was particularly apposite:

"We trained hard—but it seemed that every time we were beginning to form up into teams we were reorganised. I was to learn later in life that we tend to meet any new situation by reorganising, and what a wonderful method it can be for creating the illusion of progress while actually producing confusion, inefficiency, and demoralisation."

I think that "demoralisation" is particularly true because the process of a reorganisation always involved the threat of redundancies and the phrase "you will have to apply for your own job" – how silly that sounds.

The time for us to move from Peterborough District to King's Cross Division came along. There was already a shortage of staff in the King's Cross office so redundancy was not a problem and travel arrangements were agreed which did not leave us out of pocket but did mean much longer days. I looked forward to it. I was about 19 by then and I firmly set my sights on the bright lights of London. However, a number of staff did leave. One or two were nearing retiring age anyway. The worst affected were the staff who were already travelling to Peterborough from Boston, the victims of a previous reorganisation. Travel from Boston to London was not practical on a daily basis so they left the industry and amongst those was Derek Smith. Tim Green also found himself a promotion to a job elsewhere, so I was the only member of the bridge section to go to London.

Our new office was situated on the west side of the main station, overlooking what was then Platform 10 but is now Platform number 8. If you go to King's Cross today and visit the Parcels Yard Pub, you will be very close to my old office. The Chief Civil Engineers' offices were also located at King's Cross but on the east side of the line. These were much bigger and had their own canteen and other facilities which we could use; access being across the bridge that still spans the platforms. Whilst the reorganisation was going on, most of the staff of the King's Cross district office had successfully applied for jobs with the Chief Civil Engineer rather than waiting to see what happened in the reorganisation, so that when we arrived there were no staff at all in the bridge section and for 6 months, until more staff could be recruited, I was the only member of that section and I loved it! I got no extra money, but I was my own boss, left alone to get on with whatever I could. Furthermore, there was a bridge reconstruction in the pipeline for me to attend to which I have described in the next chapter.

The journey to King's Cross was straightforward but the morning train in particular was slow and there were very few commuters from Peterborough to London. The train stopped at several stations between Peterborough and Stevenage before running fast to London. Coming home in the evening was better. Our hours were adjusted with a nominally shorter dinner break to allow us to catch the 17.05 train which ran non-stop to Peterborough and carried a buffet car which was particularly popular on Friday evenings. Of course, working in London brought with it endless possibilities for out-of-hours activities. Generally, this meant staying in a pub with workmates and travelling home mid-evening. However, whilst I was on my first-ever lads' holiday to the Costa Brava, I met with a girl from London and our relationship continued when we got back to England. Her father was a docker and they lived in Canning Town. That was not a difficult journey from King's Cross by tube, but it did mean plenty of late nights before I could get back home. The normal journey was to catch a bus from the end of her road at about 23.30 to Stratford to be in time for the last tube into town and then on to King's Cross for the 1.15 train to Peterborough which arrived sometime about 3am, then swiftly home for about 2 hours sleep before I was up again for our 6.30 train to London. If I missed the train or just didn't fancy the late journey, I would sleep in the office and on one occasion, I managed to wangle myself into Ilford Engineman's hostel for a night.

Once the bridge section got fully manned, it was decided, much to my disappointment, that I needed some time in the planning department. I regarded this as a bit of a come down after having run the bridge section single handed but I suspect my superiors thought that I was getting a bit above myself. In truth, it was a bit of a boring few months that I spent there but there was one incident which certainly livened up the routine and got me into a bit of trouble. Around lunchtime one day, a rail broke in the switch and crossing work at Hitchin, Cambridge Junction. This is where trains to Cambridge leave the main line and it was probably the most important junction on the division south of Peterborough. No spare was available on site, but a new section was already on order from Taylor Brothers at Sandiacre, near the large marshalling yards at Toton, Derbyshire. They were able to immediately load the piece onto a bogie bolster wagon and arrange for it to be shunted out into Toton yard. I was given the job of "wagon chasing" to make sure that it was moved as quickly as possible to Hitchin.

Inter-regional communications were, at that time, very difficult, involving going through several telephone exchanges and I tried for 3 hours that afternoon to try to get through to the yard at Toton to move things along. At

4 o'clock, I gave up and decided that the only way I would get anything done would be to go to Sandiacre myself. I caught a train to Grantham, then on to Nottingham and finally a local train to Sandiacre (the station was to close not long after this) where I arrived at about 8pm. I eventually found the yard supervisor in his cabin, and he was able to locate the right wagon for me and to put in motion the arrangements to move the wagon down to Hitchin as quickly as possible. From then on, I was really only a passenger in the system, but we all carried a brake-van pass which allowed us to travel with the guard on both passenger and freight trains.

The first leg of the journey was the short distance from Toton to Colwick yard. My wagon was shunted into a train of vans and coal wagons behind a pair of Class 20 locos. The friendly driver said I would be much more comfortable with him, so I was able to travel "up front". Strangely enough, I don't remember there being a "second man" in the cab but I don't think single manning had been introduced by then so there must have been. We set off but after a couple of miles we were confronted by a red signal at Attenborough Junction. The driver rang the signaller from the Signal Post Telephone and apparently a preceding passenger train had hit several loose horses that had got onto the line, and they were waiting for a vet and the knackers' men to come to clear the line. We had to wait for at least 2 hours before traffic got on the move again and then we were low priority behind trains that were stacked up on the main line. Eventually, we got to Colwick, and I dropped off into the yard to see about an onward service. I needn't have worried, the Supervisor at Toton had done his job well, a freight bound for Ferme Park Yard in North London had been held and my wagon was quickly shunted onto it. This time, I wasn't invited onto the loco, so it was a long and rather rough ride in the brake van for me. The service was a 7* freight which means had a "head" of vacuum brake fitted wagons behind the engine to provide sufficient braking power for the train to run at a fair speed. However, this left the brake van at the back of a long string of un-braked wagons so there was a lot of "surging" and rough riding felt as we clattered down the Stoke bank towards Essendine.

The train made good progress and at about 7am, we drew up into the sidings at Hitchin Station, where the rails were offloaded and immediately used to replace the broken set of switches. I felt very pleased with myself and carried on to King's Cross by passenger train, eager to report the results of my efforts to the boss. Unfortunately, he was not at all happy and said there was no way that I should have spent all night running up and down the country. I was a bit miffed at that but despite what he thought, I was rather proud of the results of my efforts. Later in my career, when I had more experience, I realised that I could have achieved the same results by contacting "Control" who would have been able to speed the wagon to us. However, that would have robbed me of a fun night and a good story.

By now, I felt that I was due a promotion, so I started looking on the weekly vacancy list. Eventually I spotted a position for a P&T "B" grade in the bridge section at Leicester and after a successful interview I was appointed to start there in September 1968.

Chapter 5

Steam Breakdown Cranes and the Bridge at Harringay

Steam Breakdown cranes are magnificent machines that snort smoke, steam and fire. They are noisy and vibrate so much when they are working that it is a wonder that they don't shake to pieces. They can lift incredible loads but can place them with great care and precision and move them along the line under their own power or, for the heavier lifts, they can be blocked down to make them incredibly stable. Driving them is an art performed by experienced people with soot-blackened faces who work with no modern safety devices to help them.

Until around 1980, the choice of crane to handle the heavy lifts of a bridge reconstruction rested between a rail-mounted steam breakdown crane, a derrick or a crawler crane. Telescopic road mobile cranes of the type seen on the roads today had not yet been developed. Derricks take a long while to erect and crawler cranes are only really of use in open countryside so in effect, for our sort of work in 1964, the steam crane was the obvious (and in many cases the only) choice. Most Loco sheds were allocated at least one crane, and these varied in lifting capacity from 36 tons to 75 tons. Generally speaking, the more important the shed, the bigger the crane.

I first got to learn a bit about using the cranes and the protocols to be followed by listening to a conversation between Tim Green and Derek Smith soon after I started at Peterborough. They had recently been in charge of taking down the old M&GN bridge at Little Bytham, which had become redundant when the line from the Midlands to East Anglia closed. Because of the height of the structure above the GN main line, they used a pair of 75-ton breakdown cranes working in tandem. They showed me photos of the lifts and I wished I had joined sooner, just to be able to watch the operation. Unfortunately, New England shed only had a 45-ton crane and Tim had determined that its jib was too short. He therefore booked 75-ton cranes from

Doncaster and King's Cross for the job. The breakdown supervisor was, apparently, incandescent at this and raised massive objections to "foreign" cranes coming onto his patch. He would not be convinced that his crane couldn't do the job. He may well have been right because my later experience with crane supervisors was that they could always find a way. The crane supervisor was totally responsible for all lifts and at a time when there were no safety systems on the cranes to avoid exceeding the lifting capacity, their skill and experience was essential.

Every Breakdown crane came to site with a set of bogie vehicles, normally at least 2, the "Tool van" for all the chains, slings and jacks that were required and the "Riding Van" for the crew that manned the crane. Bearing in mind that a breakdown crane could be out for hours and even days when attending a derailment, the riding van carried a kitchen, a cook and some rudimentary facilities for the crew to take a rest. The second most important man on the crane was the driver who worked all the controls of the crane and I got to know a good number of them over the years. I was told never to enter the riding van unless invited and to take off my boots. The supervisor had his own office at the end of the van and in addition to the polished lino, his floor was usually carpeted.

I came into contact with the New England Crane while I was on Peterborough district and did the work on the underbridge reconstruction near Tallington. The sheer raw power of the crane was incredible and to see it working was an amazing experience. Once it arrived on site the outriggers on the crane were jacked out and down by hand onto cribs made of hardwood. The cranes jib could then be raised ready for carrying out the lifts. Huge concrete beams were lowered into position incredibly gently and with perfect precision by the driver who really operated blind and who was responding to commands and hand signals from the supervisor. I loved working with the cranes but on that bridge, I was not closely involved with the planning of the work or with liaising with the crew. It was a different matter when I moved to London, and I was told that I was to be responsible for reconstructing a part of the bridge that carried the GN main line over the Gospel Oak to Barking line just south of Harringay Station.

The preliminary works were underway on site and a new brick pier was being built on the north side of the line to increase the number of spans. The work was being carried out by the local works staff under the overall control of Jack Kilsby, the Chief Works Supervisor. He was a fiery character who sported an artificial leg. How he lost the leg, I never found out, but it never stopped him doing his job. The new concrete beams for the bridge had already been delivered and were stored in Hitchin Stockyard. Possessions had also been booked and the "day of the race" was not far away when I took over the job. I contacted the breakdown Supervisor at Finsbury Park depot and booked the crane for all 4 possessions and also for loading the beams at Hitchin each week. By this time there was no steam traction operating south of Peterborough and the new Diesel Depot had been built at Finsbury Park had been built to replace King's Cross "Top Shed". Despite this, the 75-ton crane remained steam operated for many years before being converted to diesel power.

The loading of the wagons in the stockyard proceeded without any problems, the crane crew treating it as a day out from working as fitters on the shed. Once the lifts had been done, I was invited to join them in the riding van for a full cooked meal before we returned. The actual reconstruction should have been relatively straightforward. There was no great weight in the old structure to be lifted out and the concrete beams to be installed were plain reinforced concrete without pre-stressing, so they only had to be placed onto their bearings in the right place. Of course, in the middle of the night problems will arise. The work was carried out in the winter and the first night was very cold. After removing the old structure, the preparation of the abutment tops required breaking out some of the brickwork using compressed air jackhammers. Compressed air tools freeze up in cold conditions so a fire had to be maintained on site so the tools could be de-frosted but despite this progress was slow. It certainly was not good enough for Jack Kilsby and despite his false leg, he jumped down on to the abutment top, snatched a jack hammer from a labourer and proceeded to break down the brickwork almost single handed.

The first weekend, the placing of the beams went well under the control of a young crane supervisor. He and I got on well and it was from him that I learned the hand-signals and commands to be used when carrying out lifts. Once the beams were down, a waterproofing layer was needed for the bridge. This part of the job was frequently the one that produced the most problems. On almost any bridge I have done, a different waterproofing system was specified and, in this case, the black coating to be applied over the whole bridge was found to have thickened in the extreme cold and would not flow. The cans were, therefore placed on the fire before being opened. We realised that was a mistake when the first explosion occurred. After that, the heating up was done after removing the lids.

At that time, the last passenger train from Peterborough on a Saturday evening left fairly early and I would have had about 4 hours to kick my heels in London before starting work. However, there was a scheduled empty stock train that left Eastfield yard in Peterborough at about 9.30pm and armed with my brake van pass, I used that to get me to site. The stock was to go to a siding outside of King's Cross and on the first trip I was dropped off at Finsbury Park to catch a suburban train the one stop back to Harringay. On the second week, I went down to the yard and located the train and the guard. This week it was quite a long train with, unusually, a Deltic at the head and the last vehicle was a DMU unit on its way south following repairs. The guard suggested that I would be most comfortable in the back cab of the DMU rather than in the brake van with him and I looked forward to a pleasant journey through the night.

As the time for departure approached, there was no sign of the driver or second man. The guard confided in me that the crew were from Finsbury Park depot and had brought a train north that arrived in Peterborough before 6pm. It was then the custom for them to repair to the Railwaymen's Social Club until the return journey. About 15 minutes after booked departure, we saw two men walking towards the train in a slightly unsteady fashion. They greeted us in a friendly manner and asked where I would like to be dropped off. I said Harringay was my destination, but that Finsbury Park would be fine. I got back on the train and made myself comfortable before we set off like a rocket. Using the immense power of the Deltic, the driver was determined to make up time and the coach that I was in at the end of the train swung wildly as we passed over the points out of the yard and on to the main line. Fortunately, we ran all the way on the main line under clear signals so although I am convinced that we were well over the permitted speed at places like the 60-mph restriction at Offord, the journey passed safely, if not without giving me a bit of bruising. We passed Horsey at full speed, so I resigned myself to having to make my way back to site from Finsbury Park or even the carriage sidings but approaching Harringay, all brakes were suddenly applied, and we shuddered to a halt in the station. I quickly got out, gave a hand signal which the driver must have seen because the train immediately started off again. My first job was then to go up and see the "Bobby" in the signal box to check if arrangements for the possession were all OK. Before I opened my mouth, he said to me; "Did you see that train? I have no idea why he stopped." I sensibly kept quiet.

Things were going well with the bridge reconstruction but on the third weekend, just as the crane was lowering the jib to reach out with a concrete beam and place it on its bearings, the driver hit a problem. He could not stop the jib from lowering, hard as he tried, he could not engage the gear that would have enabled his to get control of the operation and I still have the vivid image in my mind of him slamming a huge lever forward time after time only for a grinding of gears to be heard as the lever refused to engage. It all seemed to be in slow motion, but I knew that if the jib could not be held, the result would be the overturning of the crane and I shouted for all the men to move away from the crane.

The crane supervisor was standing on the ground next to the driver who was still working to get control. Breakdown cranes are fitted with clamps that can be attached to the rails to prevent movement during lifting operations and as I watched, I saw that the crane was beginning to tip forward and the rails behind it were being lifted into the air by the clamps. Only the weight of the track was preventing the crane from overturning. Just then, the leading end of the concrete beam hit the new pier of the bridge with a bang and the jib of the crane, temporarily released from the load, jerked backwards, the resulting slack in the gears allowed the driver to engage his lever and stop any further movement of the jib. Immediate overturning was thus avoided but the crane was still in a precarious position, and it took all the skill of the driver and the supervisor to coax the jib back up and return the crane to stability. After that, the remainder of the beams were installed without incident but I had to arrange for the attendance of a P-Way maintenance team to repair the damaged track before we could give up possession. I have no idea what went wrong with the operation or whether it was a mechanical failure or driver error, and I was sensible enough not to pursue the matter. Railway staff generally trusted their colleagues to deal with matters in the right way and without fuss. In later days, such an incident would have been the subject of an inquiry and probably the driver would have been suspended and the crane taken out of service until the inquiry was completed.

My shift on the bridge generally finished at Sunday lunchtime when all the bridgework was complete and I could leave my P-Way colleagues to reinstate the track and re-open the line. On the last weekend, to celebrate the successful completion of the work, the bricklayers, labourers and other staff that had worked on the bridge, invited me to a local pub for a celebratory drink. They were a great bunch of lads from many different backgrounds, including a fair number of first- or second-generation immigrants from the West Indies. I was bought a fair number of pints and felt I had arrived as a real railwayman on that day.

After that job, I had no further contact with the Finsbury Park Crane but when I moved over to the Midland Region, I was soon needing the use of the Toton based Breakdown Cranes for bridgework and I found that they were as keen as myself to do a good job and to preserve their use on civils work when heavier road cranes became available. For most of the time, Toton had 2 cranes available, the mighty 75-ton capacity Cowans Sheldon Crane (previously at Derby) and another, older 36-ton capacity Cowans Sheldon crane that had been built for the LNER and had previously been allocated to Colwick Shed. This crane was forever known as "Charlie Prince's Crane" after the legendary breakdown foreman at Colwick. I never met him, but I did work with his nephew for a while. The use of these cranes on major bridge reconstructions forms later chapters in this book but they were also used on several other types of work.

I had struck up a good relationship with the senior crane foreman at Toton and we both determined to fight off the use of road cranes wherever possible. Breakdown staff had a bit of a reputation for always having a break for dinner when you most needed them to work. I have to say that I never had a problem with this, but it was true that there was sometimes little attempt to co-ordinate the requirements of the engineers with the need for the crane staff to take a break whereas the engineer's dept. staff were only sent for a break when they could be spared. Because of this, one of the first things we did was to plan the works and the breaks so that we didn't delay the work. Some of my superiors also believed that road cranes would be much cheaper to use, and this was a major threat because at that time, the railways were so short of money that savings were always being looked for wherever they could be found. To combat this, we were able to show substantial savings, firstly by agreeing that the crane and the materials would come to site in a single train, thus saving the cost of a second loco just to bring the breakdown crane. Once on site, the crane could be separated from the materials train and use its own power to move about when necessary. We did meet with one problem when a guard refused to take the full train out of the sidings because he said, in effect, that the crane crew were passengers and could not travel on a freight train. After that, the crew had to travel to site in a road vehicle, which did mean that they could leave the depot a bit later than if they went on the ballast train.

The other major saving was achieved by looking at the overall requirements for labour on the site and including all of the crane staff in the calculations. A crane would arrive with a complement of maybe 10 staff who were needed for slinging and building cribs for the outriggers, etc., but who were underused for most of the time. However, these men agreed that they would do any work that they were qualified to carry out and since they were all trained in oxyacetylene cutting, they were very useful in the early stages of the work when the old bridges were being dismantled. In fact, on some reconstructions, the crane crew took charge of the removal of the old bridge and the civils staff signed on later. Despite this, I always had a battle to ensure that the breakdown cranes were used. I only lost the battle once and that proved to be one of the most difficult jobs I had. It also reminded me that when you have a road crane on site, all you get is a crane and a driver whereas with a rail crane you got a hugely experienced crew that do all they could to help out when problems arose.

When I was in the building section at Nottingham, I had the job of designing and building a new fuel storage facility at Eastcroft depot in Nottingham. To install the new fuel tank, I brought in the 36-ton crane from Toton. Unfortunately, it derailed on the tight curve going into the sidings and the breakdown crew had the job of re-railing their own crane before the tank could be installed. This was just one of several smaller jobs that involved breakdown cranes. On another occasion, this time when I was working in Leicester, I had the job of removing 2 redundant bridges. The first was a long single-span bridge across the Midland Main Line between Kettering and Wellingborough, known as KI Bridge because of its association with the Kettering Iron Company. The second was a small footbridge near Kelmarsh on the Northampton to Market Harborough Line. I had deliberately arranged for both bridges to be removed on the same shift in order to save costs.

The first bridge at KI was high up in the air and I determined that I needed the 75-ton crane, not for the weight but because of its greater jib length. The date for the work was Sunday 7 January 1968, but the day before, on 6 January, a major train accident occurred at Hixon when a West Coast Main Line express hit a 120-ton transformer being moved on a lorry that was passing over an AHB level crossing causing a derailment with loss of life and our crane had been sent there. Fred Spreadbury, our Chief Works Supervisor got a call from control, telling him all this and offering a 36-ton crane as an alternative. Fred asked me if we could still do the job with the smaller crane and we both agreed to give it a try. In the event, both bridges were taken down satisfactorily although the KI bridge had to be shackled directly onto the hook of the crane rather than using lifting chains. As a postscript to that tale, the 75-ton crane actually turned over at the crash site when its outriggers sank into soft, thawing ground.

On another occasion, I was at my drawing board in the office when I got a call from the local drainage authority. They had been dredging the River Leen at Bestwood Park, to the north of Nottingham when their excavator had sunk into a soft patch of the riverbed at a point where there was a bridge that carried the A611 over both the river and the railway. Their request was for permission to put a crane on the road bridge to lift the excavator out of the river. I thought that permission for this was unlikely, but I agreed to check on the possibility. I found that the bridge was the responsibility of the County Council so I gave them the necessary contact numbers and thought that would be the end of the matter. However, within a few minutes, they were back on the phone and told me that the County Council had refused to consider a crane working on their bridge, so they wanted to look at access over the railway.

Ever keen on getting work for my friends at Toton, I offered the use of a breakdown crane to pull the excavator out of the river and carry it to a point where it could get access to the road. They jumped at the chance, but I have to say, they were surprised when a steam crane arrived to do the work and I had a few choice words with their engineer when he chose to suggest that it was a bit old-fashioned. The lift was not an easy one because a large excavator is not designed to be lifted without first removing the jib; however, it was eventually done to the satisfaction of everyone. Bestwood Park was the site of sidings for serving local pits, including Calverton. It carried no passenger traffic at the time but plenty of coal. Nowadays, there is no coal but there is a half-hourly passenger service from Nottingham to Mansfield with alternate trains going through to Worksop.

This was the second time I had to arrange for a breakdown crane to go there. The first time, it was to lift in a large "Portakabin" into place. In 1973, BR introduced the "TOPS" system for the computerised tracking and directing of all its freight traffic. That meant the installation of terminals for the input of wagon numbers at many marshalling yards and sidings, especially in the coalfields. To accommodate these, unless there was already a suitable building available, large portable buildings were brought in from "Portakabin" in York by a road vehicle. I was in the building section at the time and had the job of arranging for concrete bases to be built and for the placing of the cabins. With the best will in the world, I couldn't normally justify the use of a breakdown crane but at Bestwood Park there was no alternative because the chosen location for the cabin had no road access.

Toton 36-ton steam crane rescues an excavator from the River Leen at Bestwood Park

Eventually there was a modernisation programme for Breakdown Cranes. Many of the older, smaller cranes were withdrawn and although many were scrapped, not a few remain in service with preserved railways. The more modern 75-ton cranes (now rated 76 Tonnes) were converted to Diesel-Hydraulic, and several new telescopic cranes were introduced.

I was once talking to a crane supervisor from Toton who told me that they had just lifted a Class 45 loco bodily back onto the rails using their 2 cranes (75-ton and 36-ton). A class 45 weighs somewhere about 130 tonnes so a quick bit of maths will tell you that the 36-ton crane was well overloaded but apparently it did the job without any problems and in an emergency, crane supervisors used their own judgement. Fortunately, serious derailments are few and far between these days but when they do happen it seems that the first thing that is done is to build a new road across fields to bring in a massive road crane to clear the line, a task that would be well within the capacity of a couple of breakdown cranes. The line closure extends from being calculated in hours to days and even weeks to the severe inconvenience to the travelling public and freight customers. I fear this may be because the old skills available to my generation of railwaymen have now been lost.

Chapter 6

The Leicester Experience

By the middle of 1967, I was starting to think about moving on. In fact, I had been casually looking at the vacancy list for a few months before that, but I had held back because I was in a relationship with a lovely girl who lived in London. However, that was beginning to cool so I thought it was time for a move. I had also become somewhat dissatisfied with the set-up at King's Cross. I was still enjoying the weekend site shifts, particularly track relaying but back in the office, new staff brought in to fill senior vacancies didn't seem to value the work I had already done, and I wanted to get back to real engineering, not spend my time in the planning section. Added to that I had an urge to "spread my wings" and leave home and the possibility of a promotion in grade meant that I decided that I really needed a move.

As luck would have it, a vacancy for a P&T "B" in the works section of the Area Civil Engineer, Leicester, caught my eye. P&T stood for "Professional and Technical" and there were three grades: A was the lowest and generally regarded as a starter grade whilst the majority of P&T staff were in the B grade and a few more senior staff were grade C. Later, during one or our many re-organisations, these grades were replaced by "Technical Assistant", "Senior Technical Assistant", and "Principal Technical Assistant", but they were essentially the same and eventually the "Assistant" was replaced by "Officer" (or was it the other way round). Again, there was no change to the seniority or duties, and they might as well have remained the same.

Anyway, I got an interview for the job at Leicester which I really felt would suit me in so many ways and I was happy to be appointed in September 1967. At the time, a couple of years previously, when Peterborough District was abolished, the Midland Region had a similar reorganisation and Derby North and Derby South Districts were replaced by Nottingham Division, which was housed in a new office block built on the site of the old Nottingham Loco. However, it was felt at the time that the southern extremities of the new division which extended to within sight of Bedford were too remote from Nottingham, so a small "Area" outpost was created at Leicester where mainly technical and supervisory staff were based. The District Engineer, Bill Thom, reported to the DCE Nottingham, an irascible Scot called Malcolm Harbottle, but below that level there was little contact or supervision from Nottingham, so apart from some administrative support, we were pretty well an independent office. Most of the staff were ex-Derby South and they resented the ex-Derby North staff at Nottingham so that added to the independent nature of the area. We had a small "Works" drawing office that comprised a single manager, David Pratt, Geoff Barker and myself as senior technical staff and two Technical Officers, Noella and Jenny. There was a similar sized office dealing with the P-Way.

My arrival in Leicester coincided with start of the academic year at Leicester University so most suitable accommodation had already been snapped-up by students. The list of rooms and flats that I got from the City Council office had a lot of crossings-out, showing that most rooms had been taken but I did find one that was located about 500 yards from the station. It was an awful place but as it was already quite late in the day when I got there, I took it in desperation. It was cold, damp and dirty so I determined to find somewhere better as soon as I could. Fortunately, within a couple of days I managed to find a decent ground floor bed-sit in an old house close to Victoria Park and only a 15-minute walk to work.

The room that I got was ground floor at the back, so it included the original house kitchen. I was happy there for 2 years and eventually made friends with some of the other residents. The elderly owner lived just round the corner and his daughter looked after his properties for him. She was pleasant enough but completely ignored the privacy requirements of the tenants. Using her own key and without knocking, she would come in to inspect the flat, on one occasion apparently oblivious to the fact that I was lying naked on the bed, madly scrabbling for a sheet to cover me. Somehow, I have the feeling that she quite enjoyed the experience, if only for the embarrassment that she caused me.

Later on, after I met Ina who I married in 1969, I fitted my own bolt to the door to make sure that she couldn't walk in on us. The gas fire and cooker were fed through a meter that took 5p pieces which the landlord's daughter regularly emptied. One day, just after she had left, I put a coin in and the collection tray fell off. This allowed me to keep using the same coin again and again which I did for a week or two but when I thought emptying day might be due again, I told the landlord. To my embarrassment, he praised my honesty in telling him, but of course I had saved myself a few pounds before I let him know.

I have mentioned that most of the staff in the office came from Derby and they caught a train home straight after work, so the usual source of making friends in a strange city through work colleagues wasn't open to me. For the first few months in the town, I was quite lonely and I got fed up with sitting on my own in the local pub, so I took to concentrating on my studies; I was attending Derby Technical College at the time completing my HNC in Civil Engineering. Up to now my academic achievements since leaving school could best be described as distinctly average but suddenly things became much clearer to me and I put this down to the time I spent working in the bed-sit without any distractions. I began to really enjoy my course and at the end of my final year, I won the British Constructional Steelwork Association prize for my work. One distraction that I did succumb to was to start watching Leicester City. At first, it was just "something to do" but, as any football follower will tell you, it is not possible to keep watching a team without becoming a fan. I have now followed "The Foxes" for over 50 years and I have had the joy of the Premiership winning season and going to Champions League matches around Europe as well as the lows of administration and relegation to League 1.

Before I moved to Leicester, I had only ever held a motorcycle license. There had been no requirement to be able to drive at King's Cross because we were able to use the train for most journeys but the office in Leicester had a small Bedford HA van and an old Leyland minibus for the use of technical staff so I had to learn. I took lessons but I got most of my practice in the HA van. Noella and I would slap the L plates on, and she would endure my driving and give me tips. With her help, I managed to pass my test at the first go although it was a much easier test in those days. However, that wasn't the end of it because I had to get a railway driving license and this required a second test, carried out at the BR vehicle centre in Loughborough. Once I had that I was entitled to drive any railway vehicle up to 7.5 tons and I made full use of it. Of course, it wasn't all plain sailing and in the early days I had a couple of minor prangs and brushes with the law which can best be put down to youthful exuberance.

Driving back to the office one day along Welford Road in a 30 zone and travelling at well over the speed limit, I passed a police motorcyclist parked on the side of the road and busy lecturing a motorist that he had pulled over. I reasoned that he was too busy to be able to deal with me and didn't slow down; how wrong I was. I caught sight of him in my rear-view mirror jumping on his bike to give chase. As it happens, I was approaching the area of Leicester where I lived and I had come to know the side roads fairly well, so I attempted to out-run him by doing a few turns down side roads; however, I couldn't shake him off and he caught up with me and flagged me down. I thought he would throw the book at me, especially because I had tried to get away from him, so as he slowly and deliberately removed his helmet, I feared the worse. Then I saw his face. He had the broadest grin and looking at the British Rail name on the van, he simply said, "If your trains went as fast as you drive that van, they would never be late, now get off and don't do it again." With that he laughed out loud and got back on his bike still chuckling at his own joke.

The vans were very light so in the winter when it snowed it was fun to choose minor country roads to get to site. Noella and I got caught out on one such occasion over in Rutland when our van refused to climb a steep hill on an untreated road and slithered to a halt. Just as we were wondering what to do a farm tractor came up behind us, indicated that he would help but instead of getting out a towrope, he simply buffered up behind us and pushed us to the top. When we got back, I expected to see broken lights and other damage at the back, but he had made just a couple of minor dents which would not be noticed once I rubbed some dirt into them.

Work at Leicester was varied and included both bridge and building work, much of it mundane but with the occasional bridge reconstruction or other larger project to get our teeth into. David ran a happy ship so generally it was a pleasure to go to work and we all got on well. Both Jenny and Noella took a full part in both office and site work. Noella was married and took me back for dinner a couple of times where I met her husband. He and I got on well enough, but I did think he cooled towards me after she went home from work one day with torn tights and mud on her dress. I hasten to add this was due to us scrambling up a railway embankment together when carrying out a survey! Jenny was similarly happy with the dirtier side of the work. I was amused to find out that while she had chosen a career as a railway engineer, her boyfriend was a window dresser in a city centre clothes store. Our office-based team was backed up by a number of outdoor staff and supervisors who were led by the Chief Works Inspector, Fred Spreadbury who taught me a lot about the practicalities of the job.

On my first day in Leicester, I was immediately allocated a small bridge reconstruction at East Langton, on the Midland Main Line between Leicester and Market Harborough. It was an underbridge carrying the line over a 'B' road and the existing superstructure was to be replaced with a 'Z' deck steel bridge. 'Z' decks are "through-deck" bridges and are a standard BR design for small to medium span structures. They consist of a pair of steel main girders connected with steel cross-girders and a reinforced concrete deck which surrounds the cross-girders. Each deck carries one track, and the flanges of the main-girders are off-set from the webs (hence "Z" decks) to maintain a standard "sixfoot" (sixfoot refers to the space between adjacent running lines which is approximately 6 feet wide). There is a design fault inherent in the structure because the main-girders flex under traffic and the concrete below the cross-girders is thin leading to cracks forming under each girder. These cracks are not structurally significant but do lead to water ingress and almost all have had to be repaired over the years and some bridges have already been replaced. The bridges are built off-site, including the casting of the concrete decks and then transported to site by train as an out-of-gauge load for lifting in. Because of the recent reorganisation, planning was a bit behind schedule and the date of the main possession for the work was uncomfortably close.

Still fresh from the "can-do" attitude instilled in me at Peterborough and during my Outward Bound course, I assured David that all would be well and set about organising the work which, of course, entailed the use of a steam breakdown crane on site. The actual possession work went very well, and the line was handed back on-time with a brand-new bridge in place, but I did encounter a problem during the preparatory work that almost caused the job to be delayed and I have to say, my own inexperience was a significant factor in this. We built the bridge on a stillage in Market Harborough yard. The stillage, built of timber sleepers was high enough for a rail wagon to be pushed underneath so that the deck could be lowered by jacks onto the wagon, thus obviating the need for craneage during the loading. All very well but for the concrete to be placed in the deck, the ready-mix lorries had to negotiate a steep timber ramp, and this proved to be impossible with the first lorry nearly falling off the ramp when it attempted. This caused a delay to a very tight schedule while a concrete pump was organised and the use of a very high early strength concrete to meet the possession date.

In the yard at Market Harborough, there a Ready-Mix concrete company, right next to where we were building the bridge decks and I had asked for them to supply the concrete. Unfortunately, the railway procurement system wasn't flexible enough for this and our concrete had to come from Leicester, about 15 miles away. That meant that while we were struggling with the ramp, other lorry loads were already en-route. Needless to say, the manager and staff of the local concrete company took great delight watching us try to deal with our problems and I was cursing because if they had supplied the concrete as I had requested, we would have involved them in the design and construction of the ramp, thus avoiding the problems. Anyway, it all got done eventually although I wasn't entirely satisfied with the finish achieved with the concrete soffit.

The supervisor at Market Harborough was Ted Shipley who was competent enough but laid-back and mild-mannered in the extreme. There was a system whereby the supervisor would request authority for minor works to be done by the issue of a works order but work on their own accommodation would normally be done by any staff who had a bit of spare time without bothering the system. The floor of Ted's office was rotten and one day, when he sat in his chair, the legs went through the rotten timbers and deposited him on the floor. Ted had apparently put in for a works order to be authorised but for some reason it had got lost in the system, so he just sat there until the floor gave way. This came to the ears of the boss, Malcom Harbottle, who was heard to say, "The bloody man just sat there on his arse until he fell through the floor." It was also well-known that Ted had a pretty poor home life and was dominated by his wife and daughters who would have little to do with him and who had banished him to a small part of the house. Possibly his own nature was at least partly to blame for this but there were also rumours that he once had an affair with one of our lady lookouts who knitted him a cardigan which he unwisely wore at home!

One early evening I was last in the office when I received a call from control asking if we could arrange to attend to some loose timbers in the level crossing at Seaton. This was a strange request because the line had closed about 18 months previously but apparently, the tracks and timber level crossing surface were still in place at Seaton Station and a member of the public had reported them to be dangerous. It was on Ted's patch so I arranged to pick him up from home so that the two of us could make it safe until a permanent solution could be found.

Ted met me at the door of his house, and I have rarely seen someone so pleased to be called out. We scurried off followed by a string of grumbles and mild abuse from his wife. Soon after the incident with the chair, Market Harborough depot was closed and Ted was sent to Kettering where his regular lorry driver was Brian Garner. Although Brian was a trained bridge examiner and underwater inspection diver, his regular job was driving a tipper lorry that carried materials around the worksites. Brian knew all about Ted's problems with his wife and he was also aware that Ted was a keen gardener, so one day he mentioned to Ted that he could get him some good farmyard manure from a farmer in exchange for some used timber sleepers and that this exchange would ensure the continued co-operation of the farmer in giving railway staff access over his land. Ted agreed and it was arranged that Brian would drop the manure off at Ted's house on his way back to the Leicester depot.

Brian was loaded up with a full load of very smelly and very sloppy manure to which he added an old and rusty wheelbarrow with a flat tyre and an old shovel. He took the load to Ted's house, tipped it across his drive, then knocked on the door which was opened by Ted's wife. Brian showed her the manure, the wheelbarrow, and the shovel and said, "Ted says it doesn't matter if you don't get it all shifted round the back tonight, just do as much as you can." He then high-tailed it to his lorry and away. There is no record of what Ted's wife said to him when he got home that night and when Brian saw him the next day, Ted just said, "Thank you for the manure, Brian."

The next bridge reconstructions to do were the twin bridges of Barkby Lane and Barkby Thorpe Lane which crossed over the railway just south of Syston. The bridges were of similar size and construction and because of their condition, both were scheduled for reconstruction using reinforced concrete beams. Geoff Barker took one and I took the other and I think that we both made a pretty good job of them. The actual possession work was not particularly fraught and by blocking 2 lines and having "between trains" protection on the other two lines we were able to keep traffic moving. The tightest part of the schedule proved to be the overall length of the road closure for the works which we agreed to limit to about 8 weeks for each bridge. With service diversions taking a good part of this time to complete, it was a race against time to get the resurfacing done and the roads open but both bridges were finished on time.

On the Saturday night when Geoff's bridge was being reconstructed, I had a job of my own but as it was only a short shift, I agreed to take over from Geoff on the Sunday morning to finish his bridge off. My job was the demolition of Syston Station which had just been closed and this included the removal of the footbridge across the main lines. Unfortunately, I managed to fall off the wall of the station and land on my elbow while trying to fix lifting slings onto the bridge. I was taken to A&E at Leicester Infirmary who found that I had fractured my elbow, but the only treatment was to put my arm in a sling, so I then went back to work. The station demolition had been finished in my absence, so I relieved Geoff on the bridge job where I was using a surveyor's level to set the concrete beams to the right height when Malcome Harbottle came on site. I confess that I was hoping for some words of encouragement or sympathy from the great man but all he said was, "Can you use the level with that arm," and when I said yes, he just turned away.

Syston was one of several stations closed and demolished while I was at Leicester and the policy was to get them down as quickly as possible before the locals had to chance to raise more objections and keep them open. In the case of Croft Station, on the Line from Leicester to Birmingham, the locals were so incensed by our actions that they stood on the bridge and threw bricks at us until the police arrived. The only other contact that I can recall with Malcome Harbottle was when he came into the drawing office one day with Bill Thom who was telling him about the problem we had with a road overbridge adjacent to the British Steel Works at Corby. The bridge spanned both main lines and sidings into the steel works. The soffit of the brick arch was only just above structure gauge and one of the sidings was used by huge British Steel "torpedo" wagons, which carried molten iron from one part of the complex to another. The molten steel would begin to solidify on the upper parts of the wagon to form a hard crust which then hit the bricks of the arch as the wagon passed through the bridge, causing damage.

Representations to British Steel had failed to stop the problem so I had been tasked with finding a solution. Malcome simply said to me, "What you need, Brian, is a "gabbart wi' riving knives." I had no idea what he was talking about, but I just said, "Certainly, sir," and then set about trying to find what he meant. In the end, Bill Thom had to ask him, tactfully what he meant, and it transpired what he expected me to do was to erect a steel structure in front of the bridge and fit it with knives that would knock off the solidified metal off the wagons before it passed through the bridge. Unfortunately, they would also have sliced the head off any shunting engine driver that leaned out of his cab, so we quietly forgot his idea and found a less dangerous one ourselves.

The Leicester office was always considered to be a bit of an anachronism, being small and not fitting in with the usual arrangement for Divisional Offices and it was eventually decided to down-grade it to an Area Office which meant it would no longer have any responsibility for structures so, along with the rest of the works office, I was transferred to Nottingham in 1971. However, before this happened, I had been looking after the high-level roof at Leicester Station which spanned Platform 2 and 3 and the centre roads. The 1970s was a bad time, financially, for British Rail, and the small budget available to the engineering department was barely enough to keep track and structures safe with nothing left over to spend on stations. Leicester Midland Station was in a decrepit state with uneven platforms and an overall roof that was becoming dangerous. The glazing in the roof had been removed for safety during WW2 and partially replaced with sheeting. It was now in a very poor condition with extensive corrosion throughout. I was regularly clambering over the girders checking on condition. Fortunately, because the heavy glazing had been removed, even in its poor state, it was not in danger of immediately falling down, but pieces did drop off onto the platforms below and I was very concerned that we could have a very nasty accident. It took a lot of pleading with the Chief Civil Engineer to get some money and then the funding was only sufficient to cover the removal of the roof, the laying of in-situ concrete paving over the worst parts of the platforms, and the erection of a very basic low-level awning. Contrast this with the huge amounts of money spent on stations today.

Anyway, my transfer to Nottingham was postponed slightly so that I could look after the work. First job was the concrete slab paving, which was also to act as a foundation for the temporary works being used in the demolition of the roof. Excavations were carried out using diggers brought in by rail and spoil was removed in the same way but in order that concrete could be delivered during normal working hours without disruption to trains, I decided that it would be pumped in. A concrete pump was positioned at the north end of the station in what is now the car park, and the boom of the pump was erected over the lines to the platform. So that trains could safely pass under the boom, it was secured to a signal gantry. A pipeline along the platform then carried the concrete to where it was to be laid. All went well until the time came at the end of the first day of pumping to clean out the pipeline. This is done by inserting a plug at the pump end which is then forced through the pipeline with compressed air. We had dug a small sump at the end of the line, and we optimistically pointed the last flexible length of pipe into the hole while waiting for the plug to arrive. Concrete was pushed out from the pipe into the hole, but pressure built up as the plug approached the hole and just at the last minute, the pipe wrenched itself from the grip of the person holding it and the plug came out with a bang and shot out through the roof and over the platforms to the sidings beyond. Unfortunately, all of this within clear site of passengers on the opposite platform. No one was hurt and there were no further repercussions but after that we reduced the length of the pipeline and used dumper trucks to convey the concrete along the platform. Once the new concrete had hardened, we erected heavy-duty scaffolding to support the weight of the old roof while it was demolished and then put up the ugly metal awnings which stayed in place for many years until money was found the do a proper refurbishment of the station.

I really enjoyed my time at Leicester and as well as the work, I enjoyed the camaraderie in the office and the friendships that I developed outside. It was while I was at Leicester that I joined the British Rail Diving Team for structures examinations which I have given a separate chapter to later in the book. In 1969 I went to the cup final at Wembley to see Leicester lose to Manchester City and I also got married to Ina. We rented a small flat quite near her parents, which was fortunate because the flat didn't have a bathroom, so we had to go to their house for that. Ina got pregnant with our first son, Tim, so before he was born, we bought a nice yellow plastic bath for him. Ina tried to use it herself, in front of the fire, which was a mistake and resulted in it breaking and bath water flooding over the carpet. I had been saving the deposit for a mortgage so when my transfer to Nottingham came through, I was able to purchase a house in Bingham and we moved there in 1971.

Chapter 7

Nottingham, Furlong House Drawing Office

I worked in Furlong House for 18 years (1971–1989), and until 1984, I was in the drawing office on the top floor – room 300. It was the longest period of my career in a single office and accounted for over half of my time with British Rail. During that time, I started as a Senior Technical Officer in the building section, moved on to the bridge Section as a PTO and eventually became the Works Production Assistant, which was Management Grade 1 and put me in charge of the whole site workforce as well as the works study staff in the office. I lived in Bingham and commuted to work by train. My first "boss" was John Thompson, who was in charge of the building section and also lived in Bingham. Socially we got on well and both of us played dominoes for local pub teams, but we never did hit it off at work. I was used to getting on with my work without interference from above which I always resented whereas John was never happier than when he was meddling. His nickname was "spanner man" because we all felt that he his greatest skill was putting a spanner in the works. Towards the end of my time in Nottingham, the animosity between us had grown to the extent that it was one of the reasons why I felt that I must move on. Strangely enough, outside of work we remained friends.

One of the jobs that illustrated my relationship with John Thompson was at Burton-on-Trent station. If you visit there today, you will find a rather strange combination of steps and ramps on the Island Platform. This had to be built because one night an out-of-gauge load struck the platform copings and subsequent measurements revealed that parts of the platform were 150mm too high. This instigated a rush job to restore the platform to correct levels which was allocated to me. As a part of the job, I agreed with the station supervisor to create a garden area at the Birmingham end of the platform which would be maintained by a member of the station staff who was a keen gardener. We cordoned off the area with some lengths of chain and for years it was kept in good condition. John wanted the whole area paved over because he argued that once the "gardener" retired, the land would revert to an unsightly scrub. John was partially correct because it eventually reverted to a "wild" area but one that was environmentally friendly. Today it remains a generally "wild" area of vegetation, but some maintenance work is done, and it carries advertisements for the National Forest.

My greatest love was bridges so the years spent in the building section could have been very dull but there were several projects that enlivened my time there. The biggest was undoubtedly the new station that we built at Alfreton. I was in the office in August 1972 when I was called to a meeting in Tim Green's office. Tim, my old mentor from Peterborough, was by then the Assistant Divisional Civil Engineer, Nottingham, and in his office were John Thompson and a couple of senior staff from the Divisional Manager's department. I did not normally attend such meetings, but I suspect that Tim was determined to have me involved from the start to ensure that John could not sideline me from the project. I was amazed to learn that the subject of the meeting was to build a new station. Such a thing was practically unheard at the time although the Bristol Parkway Station had opened earlier in the year. We were informed that the new station would be built at Alfreton and would be to serve the whole of the Alfreton and Mansfield area and that it would be a basic functional station with a large car park on the lines of the one already built at Bristol. At that stage no name had been given but it was later decided to call it Alfreton and Mansfield Parkway. It should be noted that at that time, Mansfield was said to be the largest town in the country without a station. John and I were given charge of getting the work done and we were told that it must open on 7 May 1973. At that stage, no detailed plans had been produced but we were given an architect's drawing showing the proposed platform lengths, general location and size of the car park.

My first job was to survey the whole site and to produce working drawings that would convert the rather sketchy detail that we had from the architect into something that could be used to build a new station. Bearing in mind that we had just over 8 months until opening day, there was no time to be lost but I did locate "The Miners' Arms" pub in South Normanton that did amazing Steak and Onion sandwiches, so that was lunchtimes sorted! The actual station buildings were to consist of a small, single-storey structure containing a waiting area, booking office and toilets. The station was being sponsored by Nottinghamshire and Derbyshire County Councils. Thankfully, they kept an arms-length watching brief on the project, but they did insist that we should utilise the "CLASP" building system that Nottinghamshire had developed for the rapid construction of schools in the county. We were only too pleased for that to happen and a local contractor who was familiar with the system took care of that work under the supervision of a Clerk of Works from the railway architects department. That left me to deal with two new platforms which were to be long enough to accommodate the Sheffield to London Expresses of the day, a large car park with approach roads from a junction with the main road and a new footbridge. All this to be designed and built in 8 months and without any major disruption to traffic because as yet no possessions for the work had been organised.

It was decided that the station would be built by British Railway's own "Direct Labour Organisation", known to everyone as the DLO. Unlike BR's own workforce, the staff in the DLO were employed under Construction Industry terms and conditions rather than the normal railway staff T&Cs. The organisation had been set up to carry out large parts of the 1955 modernisation plan on the London Midland Region, including the electrification of the West Coast Main Line and it proved itself so effective that it was continued with until the 1990s. There was some animosity between the DLO and the regular railway staff, supported by the National Union of Railwaymen and rules were in place to ensure that the DLO did not take the normal work of the railway staff, but the point was that it gave us an excellent extra in-house resource that avoided the use of contractors and the union appreciated that. The DLO was based in Birmingham, and it had a varied workforce but had very strong representation of Irishmen who had been recruited from Ireland in the '50s and '60s and a group of really excellent Sikh tradesmen from the Birmingham area. The foreman carpenter/joiner in particular was an excellent man who often shared his pack-up lunch of cold curry with me.

Needless to say, the budget for the new station was very limited and not only were there "no frills" but every opportunity to save money had to be taken. The most difficult job was to construct the platforms. These were designed as a pair of concrete block walls topped with concrete platform copings running parallel to the track and some 900mm high x 240m long. The front wall was only just to the side of the sleeper ends and the mass concrete foundations for the wall had to be built about 600mm deep. Fortunately, the one thing that BR engineers were trained and experienced to do was carrying out heavy engineering works immediately adjacent to the track. Like most of my colleagues, I had attended the excellent "Heavy Engineering" course at Watford training centre and of course the DLO staff worked alongside the railway in the normal course of their work. Nevertheless, it needed some innovative thinking and the co-operation of the operating department if the station was to open on time.

Access to the site by road was easy on the down side of the line, directly through the old station yard but access to the up side was only possible by driving through the disused bore of Alfreton Tunnel so we had to make that route suitable for access by heavy plant. We also decided, where possible, delivery of bulk materials would be by rail. The line at Alfreton was the old Erewash Valley Line which carried a normal service of one Sheffield to London Passenger train in each direction and numerous freight trains every hour, most of which were coal trains but also some steel and aggregates. The line speed was generally about 80 MPH, but it had several speed restrictions due to mining subsidence so timings over the route were not tight and we were permitted a 20 MPH speed restriction while work took place. We used 360-degree excavators and manual labour to excavate and cast the foundations working on a "between trains" arrangement which would not be possible today.

The operating department diverted some freight away from the line to assist us, sending it instead along the Derby-Leeds Line but we still had the 2 passenger trains to contend with together with at least 1 freight in each direction every hour. Normally, "between trains" working requires work to be suspended and all protection removed before trains approach the site so that they are not delayed. Such an arrangement can mean that work has to be suspended for about 15 minutes for each train that passes, and this would have severely delayed the work. Instead, a Divisional Inspector was provided on site and dispensation was given for some delay to traffic. Protection was therefore kept in place and signals were held at danger until trains were in sight, work then stopped and protection was quickly removed to enable signals to be cleared and the train to pass with the minimum of disruption. Of course, it is only possible to use such a system where line speeds are fairly slow, but it enabled the platforms to be built very quickly. Once the walls were built, they had to be filled with compacted aggregate before being surfaced with tarmacadam. This is where we made some very considerable savings. When tracks are ballast-cleaned and re-ballasted, the ballast cleaning machines remove the stone ballast from under the sleepers, pass it through a series of vibrating grids to remove as much of the dirt as possible together with any under-sized stones and then return the good stone back to the track where it is supplemented with new ballast. The material that is not re-used is known as "Spent" ballast and is loaded by conveyor belt onto rail wagons and taken away to a railway "Tip" for disposal. The project is charged a fee for the cost of unloading the wagons at the tip and for the general running of the facility. Spent ballast is an excellent fill material. It consists of hard aggregate mixed with a fair amount of dust, some oil that has dripped from trains and other detritus which helps bind it once it has been consolidated.

To avoid the cost of new stone for the platforms, we simply diverted trainloads of spent ballast directly to the site from ballast cleaning works and unloaded it into the platforms from out of the rail wagons. We managed to save quite a lot of money on the project in this way because not only did we not have to pay for the stone, but the cost of the trains was included in the cost of the track works where it had come from, so I had no ballast train costs debited to my job. Ever the optimist, I also tried to get the project credited with the money saved in not unloading the wagons at the tip but this was where my P-Way colleagues drew a line and pointed out to me that if I pursued that argument too far, their co-operation in providing the ballast might not be so forthcoming in future. Fair enough, I suppose.

The platforms were built in good time and to the specified height that prevailed at the time. This was not without some difficulty because the track was on a "transition curve" meaning that the curvature of the line and the cant of the track varied along the length of the site which in turn meant that the height and location of the platforms varied every few metres. It is also, far as I am aware, the only station to be built to the lower height of platforms that was mandated for the proposed high-speed trains (although Bristol Parkway might also have been built to the same height). The normal height was reduced from 910mm to 750mm just before we started on the work but after completion of the station, this changed was abandoned and standard platform height reverted to 910mm. The platforms, were therefore, built lower than was necessary.

Construction of the car park, erection of fences and roads generally proceeded without any great problems. We did have to slightly alter a wall at the road junction to improve sighting but otherwise there was no interference from the County Council. The footbridge was delivered to site by the manufacturers and was erected during a short Sunday morning possession. The station was opened on time by Sir Peter Parker, Chairman of BR. I was not invited to the ceremony, which was mainly attended by local councillors, but I did go to site early that morning to ensure that everything was shipshape, and "Bristol Fashion" and I stayed on in the background to watch the proceedings.

I visited the station recently and very little has changed in the 50 years since it was built. Most of the timber fencing is original with just a few palings having been replaced. But everything else seems in fair condition. Of course, it doesn't have all the bells and whistles that seem to be considered necessary for a new station today, but it took a fraction of the time to construct and cost much less. A recent station at Kenilworth seems to have taken over 10 years to build if you include the time spent on obtaining finance and planning permission! The only disappointing feature is the loss of disabled access to the up side, which we provided by a barrow crossing that has since been removed. Ramps or lifts to the footbridge are urgently needed. Few of the trains that stop there use the full length of the platforms but in 2020 there were still a couple of London trains that stop although these were under threat. The normal daytime service is provided by one Norwich-Liverpool train and one Nottingham-Leeds stopping service in each direction. The station was 9 miles by road from Mansfield itself and when Football Specials stopped there, busses were provided to take the fans to Mansfield Town's ground. However, on more than one occasion, during the years when football vandalism was at its height, fans were marched the whole 9 miles to and from the ground. In 1995, the original station in Mansfield re-opened with a direct service to Nottingham over the Robin Hood Line and my station is now just called Alfreton and is used by about 300,000 passengers per year.

Alfreton Station during construction

Alfreton Station in 2020

Most of the work for the building section was dealing with defects in station and depot buildings. I had to spend a lot of time at Toton Diesel Depot and visits there were always interesting. One job that I had was to install heavily reinforced concrete ground beams inside the depot to support the jacks that lift the locos off their bogies. Another and totally thankless job was trying to cure leaks in the large flat roof that covered one end of the shed. What was needed was for the covering on the whole roof to be renewed but we never had the money for that and spent many fruitless hours trying to find and repair individual leaks. Away from Toton, I did get a couple of good jobs in the Nottingham area. The first was the conversion of the old London Road Low Level Station that was used for all Nottingham parcels traffic to a depot suitable for handling BRUTES. This acronym stood for British Rail Universal something-or-other and was the name given to wheels cages for parcels that could be moved on and off trains and road vehicles. To do this we removed some tracks and constructed larger and higher concrete platforms.

On one day we installed over 120 cubic metres of concrete in floor slabs. Casting the concrete was finished by 5pm but a mechanical float finish was specified for the work, and this has to be carried out just at the point where the concrete is ready to set which in this case was about 2 o'clock the next morning. It took all my powers of persuasion and promises of a bonus to get the lads to stay on! The other big job at Nottingham, and one that I am particularly proud of was the re-slabbing of the platforms of Nottingham Station. If you go there today and look at the slabs that cover most of platforms 1, 3, 5 and 6 you will see that they are as solid as they were when they were installed some 50 years ago. Any loose or broken slabs will be areas that have subsequently been installed or altered. As I have said on many occasions, money was tight so the slabs were plain grey concrete, but it was essential to get the installation right because uneven slabs that cause tripping hazards are a real no-no on stations, so we installed a layer of well-consolidated stone below the new slabs. There is no road access to the island platforms at Nottingham so the main plant used was brought in by train and consisted of small diggers called "BOBCATS" that could be manoeuvred around the columns on the platform, tandem rollers and vibrating plate compaction equipment. All materials also came in by train including the graded stone aggregate base. Some of this was unloaded by excavators from standard rail wagons but just about the time that we were doing the work, I read about a new self-discharge train that had recently been introduced by Redlands at Mountsorrel, near Loughborough. The train consisted of a rake of wagons that were permanently coupled together and connected by a conveyor belt that ran under the train and led to a swinging arm discharge unit. I contacted the company to see if they would deliver stone direct to the station in their train and unload it onto the platforms. They were sceptical at first because the train was already fully utilised during the week delivering roadstone to depots around the country, but when they realised that we would be requiring the train on a Saturday night when it wasn't normally in use, then they were more than happy to oblige.

The operation was a great success with the stone being placed along the full length of the platforms to exactly where we needed it. Following this work, which I think may well have been the first use of the train on an engineering job, the self-discharge trains were regularly used on track-laying and re-ballasting sites. I was very pleased with the end result, although, true-to-form, the only comment that I got from John Thompson was a complaint that the staff on night shift had been using the station lifts and staircases to move the plant around, an initiative that I personally approved of! Before the job started, we tried very hard to get the steel-wheeled station trolleys replaced with new ones fitted with hard rubber tyres. Unfortunately, that proved to be beyond the station budget so you will see the edges of many of the slabs have been chipped by the steel wheels which detracts somewhat from the appearance but not the stability of the slabs.

Eventually, I got my move to the bridge section where I joined Alan Hughes, Pete Fisher, Geoff Barker and John Colbourne in a merry group of bridge engineers, and it was an incredibly satisfying time for me personally. I have included separate chapters dealing with bridge reconstructions and tunnel works but aside from these projects, the regular work of the section was always interesting. Repairing railway bridges always involved heavy engineering and relied primarily on traditional methods and skills. That is not to say that we were not forward-thinking when the job needed it but repairing old steel, masonry and even timber bridges required the use of matching materials and therefore the skills of trained craftsmen. Fortunately, the skills that we needed could always be found somewhere within the railway family. We had bricklayers at Derby and elsewhere who knew how to "turn" a brick arch and carpenters who could work with the heaviest of timbers. We were short of a substantial steel gang at Nottingham, ours having been whittled down over the years to just 4 men but at Sheffield, the next division to us, there remained a full outside gang which was supported by "Woodburn shops", which could manufacture all the steelwork that we ever needed. One of my favourite jobs was producing designs for bridge repairs and then being on site while the repairs were carried out.

Repairs to brick bridges were commonplace and once the extant of the repairs had been agreed on site, then the bricklaying gangs could do the work with the minimum of input from the drawing office staff. However, just occasionally the first 2 rings of a complete arch would need replacing and this required the design of large timber centres to support the arch itself when the old bricks were removed and the new bricks until the mortar had set. This was interesting work and the appearance of the new brickwork was totally dependent on the skills of the bricklayer. We did this to a side span of a bridge on the Sheet Stores to Stenson Line. The bridge was a "skew" arch which facilitated the road crossing the railway at an obtuse angle. To achieve this, the brickwork of the arch must follow the skew and is not at right angles to the piers. This is very skilful work which we entrusted to our most experience chargehand, Geoff Stevenson. We also made sure that several apprentice bricklayers joined him to see how it was done.

Geoff was nearing retirement and could tell many tales. He had worked extensively on the tunnels of North Derbyshire. One in particular, Clay Cross, had a particularly high profile so that a timber platform could be erected above the tracks that allowed repairs to be carried out whilst trains ran below. This was, of course a very economical way of working but hard on the bricklayers when the trains passing below were steam-hauled. Geoff also pointed out that access to the platform was by a ladder lowered down to track level between trains with a lookout keeping watch. Not something that would be allowed today although such practises continued well into the 1980s. For instance, when the East Coast Main Line was being electrified, much of the installation of equipment above the line was carried out from ladders under lookout supervision and I was told that to avoid going up and down ladders all the time, workers would stay clinging onto the head spans while trains passed below.

Brick arch repairs using timber centres

Repairing steel bridges generally required detailed drawings showing exact sizes of repair plates, bolts and even rivets. I carried out quite a number of these jobs and well remember carrying out the calculations with a slide rule which was, in fact, a very fast and accurate operation once one became experienced in its use, then standing at my desk detailing the work on a series of drawings while my cigarette burned out in the ashtray next to the drawing board. Several jobs stand out in my memory. One of these was the footbridge at the level crossing at Oakham, which is the county town of Rutland. The railway runs through the centre of the town and crosses the main street on the level. The signal box here will be familiar to model railway enthusiasts because it was the inspiration for the Hornby one. When the level crossing is closed, pedestrians can cross the line over a beautiful metal lattice footbridge of typical Midland Railway design. This had become severely corroded and consideration was being given to its replacement. Any new bridge would have been a heavy steel structure with solid parapets to meet modern standards and would have been completely out-of-place so I persuaded my boss that a repair could be carried out economically. The repairs required were so extensive that the bridge would have to be taken out of use for several weeks. It would have also been extremely time consuming to do the work in-situ during short possessions and the quality of the final product would also have suffered.

Having assured the local authority that closing the footbridge was only temporary, therefore, we dropped in one night and loaded the bridge onto rail wagons and carried it off to our yard in Sandiacre, leaving the good citizens of Oakham to wait at the crossing gates when trains were passing. The bridge was stripped of all its timber decking and stair treads then blast cleaned to remove all old paint and rust. We then had to replace or repair many of the structure's wrought iron elements. Normally when this is done to a heavy iron bridge, rivets are replaced with High Strength Friction Grip bolts, but these clamp together the parts of a structure in a very rigid fashion which is not considered appropriate for lighter weight lattice structures so "Black bolts" are often specified as they are not so tightly torqued-up and more resemble a riveted connection. This set me thinking about rivets which are aesthetically much better looking than rows of bolts but which had largely gone out of use in bridge work so I contacted the Steelwork Supervisor at Sheffield and asked if his gang could still do a riveting job. He had to check and get back to me but within a day-or-so he confirmed that they still had the equipment and staff with the skills to do the work. Watching them work was amazing, red-hot rivets were thrown to the riveters who drove them into place with compressed-airs hammers. The lads doing the job really threw themselves into it and confirmed my long-held belief that there was always someone in the railway family to do any job that was needed.

Once the repairs had been done, the bridge was spray painted and new timber decking fitted. The job did take longer than I had expected and the locals in Oakham were getting restless with more than one letter to the local paper alleging that the bridge was gone forever and we never intended to return it. However, we got many complements about the workmanship once we had put the bridge back up. I was in Oakham recently and the bridge still looks good, however, it has had to be raised to accommodate container traffic and the sections that have been added are of a bolted construction. Fortunately, they are at low level and not too obtrusive but they are much heavier looking, so I am glad we chose to re-rivet the main structure.

The most extensive repair that I undertook was in 1978 to a main road bridge immediately south of Derby Station which crossed 4 tracks of the Derby-Birmingham line and sidings into Derby Carriage Works in a series of wrought iron spans with brick abutments and piers. Much of the wrought iron was severely corroded with quite a few sections that had holes in them where the corrosion had gone through the full thickness of metal. It was probably touch and go whether to replace the superstructure completely or to repair the defective parts but again, money was tight, so the decision was to extend the life of the bridge with repairs. The aim of the work was to restore all parts of the structure to full-strength and the work involved repairing corroded parts of the larger girders with new steel plates and replacing many of the smaller girders with new. In order to get access to some girders, it was also necessary to break out brick jack-arches and replace them with new reinforced concrete deck slabs which I had the opportunity to design.

A series of meetings with Derby City Council resulted in an agreement to close the road for 6 weeks and alternate pavements for longer period. Possessions of the line were limited to the weekends but were of reasonable length because we could close 2 tracks out of the 4 available. The parapet girders on both sides of the bridge were in very poor condition and required repair plates to be fitted over large areas. All of these were identified, and an interesting system was evolved to carry out the work. Girder brackets were fitted to the outside of the parapets and working platforms were put in just above the top of the trains. We then arranged for the Sheffield Steel Gang to work on nights at the bridge, removing rivets and making timber templates of the repair plates needed which they would then take back with them and drop off at Woodburn shops on their way home. Woodburn would use the templates to make new plates which were brought back and fitted the next night. The smaller Nottingham gang was used for those repairs that could be carried out during the day.

The system worked very well and progress was excellent although the 6-week road closure was still very tight. The steel workers who worked on the platform were only just above the open lines and had been warned not to drop any tools but we forgot that, in order to get to the lowest parts of the structure it was necessary to reach down below the platform and I was horrified one day as I watched two men manoeuvring a steel plate into place whilst sitting down with their legs dangling down in front of trains. I was about to shout a warning when a train approached and the two calmly lifted their legs while it passed then lowered them down once it had gone.

In the days leading up to the road opening, we had let a contract to a road surfacing contractor for the final surfacing, but our own staff were used to lay in the stone sub-base over the new deck panels. We hired a large road-roller to consolidate the stone, but I had not given thought to who would drive it. Normally, there would always be someone on site who would volunteer but, on this occasion, bearing in mind that the roller had to be guided around girders and other obstacles, there were no takers. Time was of the essence so there was no choice but for me to have a go. I have to say it was great fun, but I was glad to complete the job without damaging any of the new work. The bridge was replaced a few years ago with a nice-looking steel bowstring bridge but the work that we did extend the life of the old bridge by over 35 years, so it was certainly a cost-effective repair.

Bridge 3 – Derby. The Carriage Works span

Bridge 3 – Derby. General view

Bridge 3 – Derby. Main Girder web repair

Bridge 3 – Derby. Cross-girder repairs

Bridge 3 – Derby. New Deck panels

While in the bridge section there was another bridge incident that I got involved with but this time it was Alan Hughes who got the bulk of the work, much to my chagrin, but I did manage a minor role which kept me in touch with all that was going on. An excavator on the back of a low-loader hit the bridge carrying the South Leicester Line over the A5 to the west of Hinckley. The arm of the excavator wedged under the girders and pushed the bridge up into the air. It was one of the few occasions when a road vehicle hitting a bridge didn't come off worst. Fortunately, men working at the nearby scrap yard heard the bang and were able to warn an approaching train that managed to stop before it arrived at the bridge. Had that not happened, then the train would have been derailed and almost certainly there would have been major injuries and loss of life. Alan and I were in the office when the news came in and Alan was despatched to site to see what could be done. Before he left, it was agreed that I would see if I could source some emergency bridge girders so that the bridge could be reinstated. I knew that we had one set in the yard at Sandiacre which had been recently used elsewhere on our division and had not yet been returned to the Midland Region central depot at Newton Heath near Manchester. As luck would have it, they were long enough and within a couple of hours, I had organised for a "Long Load" lorry to go to Sandiacre and the Toton Steam Breakdown Crane to load the girders. These beams would be enough to reinstate one track and restore the service, albeit with single line working and by 2pm they were loaded up and we were forecasting that the line would re-open sometime that evening and certainly in time for a passenger service to run the next day. Unfortunately, Keith Ratcliffe, the AACE (Works) then decreed that before any work was done on site, we needed authority from above and we also needed the railway photographic unit out to fully document the site in case of legal action against the lorry driver. I was fuming and I have always suspected the hand of "spannerman" in the decision. I thought it was totally against what we should be doing to delay the re-opening of the line for any reason, however, I was out manoeuvred on this occasion. Nevertheless, the plan worked well and the following day the temporary beams were installed.

In the meantime, I ascertained that Newton Heath had a second pair of girders in stock and we got those sent to site by train so that both lines could reopen. The bridge then remained as a temporary structure until a new "Box-Girder" bridge could be designed and manufactured. Both Alan and I did shifts on site when the new bridge was installed. I had to defer to Alan in the choice of crane and, to be fair, an 80-tonne road crane was the appropriate choice. However, we did deliver all the materials to site by rail and with just one ballast train. The only snag with that was the length of train which had to be moved on site. Once the temporary bridge had been removed, of course, there was a big gap in the line and to move the new bridge girders into position for lifting, the train had to be propelled to within 300mm of the end of the track. We had not yet got two-way radios for use on site, so I was pleased that we had a divisional movements inspector on site at the time and he bravely called back the train by hand signals with the loco several hundred yards away. Fortunately, both DI and driver were on the ball!

The Leicester to Burton Line was badly affected by mining throughout most of its length. Towards the Leicester end there were collieries at Ellistown, Bagworth and Snibston which were part of the Leicestershire coalfield. The seams were quite shallow and it was reputed that people living in the area could hear the miners talking below their houses. No doubt that was an exaggeration, but the shallow seams did mean that the railway in this area suffered badly from subsidence and colleagues from the Permanent Way department spent a lot of time repairing the track. The main method of doing this was to lift the sunken track until it regained a reasonable profile, but the structures sank as well and the correct relationship between track and structures had to be maintained.

One day, a ballast train was out on the line supplying new ballast whilst John, one of the technical staff from Furlong House was setting out the amount that the track was to be lifted. Work went well throughout the day but after the train had been unloaded and the track had been lifted, they found that the engine could no longer get under a small overbridge at the site. John had forgotten to ensure that the structure gauge was maintained, and it was fortunate that the line had not been opened to normal traffic in that state, otherwise there would have been a major incident. Lifting tracks is a fairly easy operation, either mechanically or by hand using jacks but there is no way of lowering tracks again without a major excavation job so we got a call in the bridge section to see if we could help. I went to site, by which time it was early afternoon and had a look at the bridge. It proved to be a farmer's occupation bridge which was only used at harvest time and that was several months away so I decided that the best way to get the line open quickly was to lift off the bridge completely so that the stranded train could be released and then prepare a scheme to put the bridge back at a higher level in time for the harvest. The farmer had no objection, but he did tell us the there was a small water main on the bridge. I requested that the water authority come to site to divert the pipe and I ordered up the Toton Crane to lift the bridge off. The crane, with a full team of men arrived before the water had been turned off so while they were waiting, the whole crane crew walked over the fields to the local Miners Welfare where I joined them for a couple of pints of bitter. By mid-evening the water main had been isolated and the crane made short work of lifting off the bridge.

I spent the long hot summer of 1976 at work while many of my colleagues were taking holidays in the sunshine. My 2 children were very young at the time, so we had decided to take our holiday in September when the older children were back at school and prices were cheaper. The hot weather also persuaded us to stay in England and we elected for Scarborough. Unfortunately, on the day that we started our holiday, the heat wave broke and we emerged from Scarborough Station in a thunderstorm. Little use was made of the swimming trunks and buckets and spades that we had brought with us but the yellow mackintoshes and sou'westers that we bought on the first day were invaluable. I still feel guilty when I think of the whole of the summer that my wife and children spent in Bingham when I should have been taking them on holiday, added to that, I was very busy that year and worked most weekends, as well as Monday to Friday.

The biggest job that I had at the time was the repair and strengthening of Mill Road Bridge in Wellingborough. The bridge is situated just north of the station and it spans the fast and slow lines. It is still there but a new overbridge has been built next to it which carries most of the traffic. I remembered the bridge well from my trainspotting days because on trips out from Peterborough, we would alight from our train at London Road Station, run up to Midland Road Station to see the Thames-Clyde express heading north and then cross Mill Road Bridge and attempt to visit Wellingborough shed. The attempt was usually successful but if not, many of the locos in the yard could be seen from the bridge. As well as access to the loco and sidings, the bridge gave access to Whitworth's depot where all sorts of foodstuffs were kept and by 1976, the lorries delivering them had grown in size and weight and the narrow bridge was showing the strain. We stabilised the east abutment with ground anchors and by cement grouting but the arch was badly fractured and before repairing the brickwork, I decided to provide a concrete saddle which would spread the wheel loads and reduce point loadings on the arches. The difficulty with this was that there was no diversionary route for Whitworth's traffic so the whole of the saddle had to be cast and the road surface reinstated over a weekend.

Physically doing the work in that time was not too much of a problem but if the concrete was not to be fractured by the first lorry crossing the bridge at 5 o'clock Monday morning, then it had to have gained 80% of its strength in about 20 hours. This is very fast so I decided on using a vacuum de-watering system which would enable high early strength to be achieved. This was duly employed but to be honest the weather was so hot on the day that we were more concerned with spraying water onto the concrete to prevent cracking. All went well in the end and a combination of the repair and strengthening work, together with providing a much smoother road surface for the lorries resulted in the bridge being stable again and it continued to support very heavy road traffic for over 40 years. Any work involving brickwork and concrete requires large quantities of water and the drought had resulted in an outright ban on the use of mains water for construction projects, so we decided to make application to withdraw water from the nearest substantial watercourse. This water was not used for the structural concrete which was supplied ready-mixed to site but it was used for the cement grouting and general site use.

I duly made application to the local drainage authority and our account's department paid the necessary fees which included the cost of advertising the application in the London Gazette. I knew little about this process but was happy to follow advice and our notice duly appeared, although we did not receive a copy ourselves, so I never actually saw it. However, someone in British Rail did because within hours of its publication, I received a very irate phone call from BRB HQ in London. Unbeknown to me, the railway employed a person whose main job was to place notices in the Gazette. Bearing in mind, the railway was by no means expanding at the time, the job was not an onerous one and I got the impression that I had robbed the gentleman concerned of one of the very few chances he had to exercise his talent. Eventually, he accepted my apology with good grace, and it was another reminder of the wealth of expertise in the railway family could have been used instead trying to do everything myself.

During my time at Leicester, I had completed my part-time studies and gained a Higher National Certificate in civil engineering (with endorsements). When I started the courses, back in 1964, this would have led to making an application to become a Chartered Engineer, but the rules had been changed over time and I had to make do with a Technician Engineer status in the Institution of Civil Engineers. I achieved this whilst I was in Nottingham and by 1984, I was feeling that I should look to moving into "management", so I eventually left the Bridge section, although I remained at Nottingham.

Chapter 8

Drinking

In common with almost all industries, there was a culture of heavy drinking in British Rail when I joined it in 1964, which of course led to many humorous incidents that I can recall. However, it also led to accidents and to alcoholism, so in the 1980s British Rail tightened the rules and became one of the first major industries to instigate a total ban on drugs and drinking with a testing regime to back it up. It was a very good thing, and it is a pity that some branches of industry, even to this day, have not adopted similar measures. There were only a few grumbles at the time, and it seemed to me that the new rules were accepted very quickly and very readily by almost all concerned. I was working at Nottingham when the ban came in and it stopped us going to the pub at lunchtimes but the only complaints that I recall were from BR staff working on Derby Station on nights who had been accustomed to having a few pints with their colleagues from the post office who loaded and unloaded the mail trains. The Royal Mail had not adopted a no-drink policy at the time so the "posties" could have a drink, but the railway porters could not, and this apparently led to some friction. The need for such measures was amply illustrated in 1972 when an excursion for railwaymen and their families crashed in South London and the train driver was found to have been drinking heavily in the railway club at Margate before taking charge of the return train.

To return to the beginning, I had my first pint in a pub when I was about 14 years old. I was in the army cadets at the time on a training weekend in Northampton and I accompanied older cadets to a back street pub where we were served pints of bitter at 1/3d a pint. We all had our uniforms on and confidently assured ourselves that the landlord would assume that we were regular soldiers. Of course he didn't, but neither was he expecting a visit from the police, so he was happy to accept our money. As the years went on, along with my mates, I became a regular drinker, particularly a local called the "New Inn" in Woodston. It was there that I celebrated my 18th birthday, much to the consternation of the landlord who had been serving me for at least 2 years. It was also there that I had a brush with the law which could have resulted in a criminal record. There are only a limited number of games that can be played in a pub for "small stakes". These include dominoes, darts and cribbage but definitely not the harder gambling games of poker and brag that we were playing in the front parlour.

One night when we left the pub, we noticed a policeman in uniform standing at the upstairs window of the shop opposite but being young and very naive, thought no more about it until one night when 4 or 5 burly policemen forced their way into the pub through doors and windows. The sergeant leading the raid immediately confiscated all the money on the table and our names and addresses were duly recorded. There were two card schools going on in the pub at the time, one for older men who were gambling for quite high stakes, and our game, which involved much smaller sums of money but was, nevertheless, illegal. Fortunately, the police were really only interested in the high stakes game and the rest of us just had to report to the police station some days later for a dressing-down. I don't think anyone saw the money again after the sergeant picked it up so I can only assume that the police benevolent fund was the beneficiary.

When I joined the railway at Peterborough, drinking "on duty" was limited to a couple of pints at lunchtime and this was generally accepted as being reasonable. The only heavy drinking that I became involved with was on trips out or at Christmas. On my first Christmas, there was a party in the office with major amounts of drink provided including spirits. I was 18 by then and joined in rather too heartily to the extent that I had to be taken home by the driver of the relaying gang bus. I think he had been warned about my mother because he half-carried me to the back door of the house, knocked loudly then beat a hasty retreat. Also, that first year, I was able to go on a Permanent Way Institution trip to Linz in Austria and then on to visit the Octoberfest in Munich. We all used our free rail travel, but the rest of the trip was paid for by Plassers, the makers of the big yellow tamping machines and the first part of the trip to Linz was to visit their factory. This gesture of goodwill was common at the time but later it got out of hand with accusations of bribery and so I never went on a similar "paid for" trip again. Having been looked after well by Plassers on the first leg of the trip and taken by them from Austria to Munich, where we were to provide our own accommodation. Unfortunately, by the time we got around to organising things, there were no hotels available for us in Munich. Instead, our counterparts in German railways organised for a "Couchette" coach (a sort of very basic sleeping car) to be stabled in a siding for our use, complete with an attendant. This was very good of them but several of us in an inebriated state from sampling the joys of the Beerfest spent some hours in the dark wandering the sidings until we found our coach. Fortunately, we didn't stray onto the main line!

I have said before that the main night to have a drink for railway engineers is a Friday, followed by a lie-in on a Saturday morning to prepare for a Saturday night shift and that is true. However, in my younger days, I didn't entirely follow that rule, especially, as often happened, when I had a shift starting in the early hours of Sunday morning. I learnt my lesson one very cold winter's night when I was booked on for de-stressing some track south of Peterborough. I went to the pub as usual that Saturday evening and then on to a drinking club until my shift started at about 2am. I did moderate my intake of alcohol but was still way over any limit that would be acceptable today. When I got to site, I was told that the cold weather was delaying the work so once I had checked that I had all the necessary equipment, I laid down on the de-stressing trolley to wait and promptly fell asleep. The next thing that I knew was being shouted at that I was needed but when I tried to get up, I found that I couldn't move. I panicked at first, thinking that I might have some sort of alcohol-related paralysis but then realised that I had frozen to the wooden deck of the trolley and had to push myself up with hands and feet to break the seal. I felt awful and vowed "never again".

Working down at King's Cross involved regular moderate lunchtime drinking for some but we had a short dinner break so that we could adjust our hours to match the train times to and from Peterborough, so this was only an occasional pleasure for me. However, I did sometimes stay for an evening's drinking, but the problem was remembering to alight from a late northbound train at Peterborough and on 2 occasions, I slept on. On the first one, I woke up as we approached Grantham but when I got off, there were no passenger trains back to Peterborough until the morning. Fortunately, I had my Brake Van pass with me, and I was accommodated on a southbound parcels train which took me to Peterborough East. On the second occasion, I woke up as we pulled into Newark Northgate but when I shouted to the station staff about a train back to Peterborough, he told me I would be better staying on to Doncaster to get a train home. He was right, and I went back on a Southbound Sleeper service that got me into Peterborough at about 4am. Just in time to get in to bed before my mother was shouting me to get up for work!

There was little drinking done while I was at Leicester, mainly because very few of us actually lived there and when I first moved to Nottingham, lunchtimes were pleasurably spent playing cards in the office. However, after a while, I did start to visit the "Loco", a small old pub near the office along with most of the bridge section and a few others. At first it was a quick pint and back in half an hour, but as with most of these things it began to get out of hand. One pint became 2 and then 3 and towards the end of the week, we could be seen creeping back to the office at 3 o'clock in the afternoon. We justified this to ourselves by the fact that we had to work difficult shifts at weekends but really, we had no excuse and eventually it stopped when we all received a letter of warning. One character in the office, a former Guardsman, had a real drinking problem. I don't know where he went at lunchtimes but at about 3 o'clock in the afternoon, he would walk through the door of the drawing office, be seen to gather himself and then march "ramrod" straight the full length of the office to his desk while his section leader, who was afraid of him, hid his head to avoid having to confront him.

Each year at Nottingham, we looked forward to the "office trip", which was organised very well by one of our mates. The favourite destination was Amsterdam, but other destinations included Paris, Hamburg, Rotterdam and even Copenhagen. Travel was always by train and involved setting off after work on Friday evening and returning on Monday morning with some dispensation allowed for us to get back to the office at about 11am. On all of the trips except Paris, travel was by catching the Harwich Boat train which left Nottingham at around 5.30pm and connected with the overnight ferry from Harwich Parkston Quay to the Hook of Holland and then onwards by early morning train to our destination. All train travel was free and only a small sum was payable for the boat to cover port taxes etc. The return working arrived back in Nottingham before 11 o'clock on Monday morning and most of us went straight to work so as not to use up a day's leave.

Of course, drinking in the buffet on the train south was the order of the day, followed by staying up in the bar as late as possible on the boat before getting a few hours rest and disembarking for breakfast. I usually stayed the course pretty well and on one memorable occasion I found myself going back to my cabin on the boat just as my companions were getting up to leave. This all failed on one trip when I was talking on the train to a lad called Steve Belasco, whose claim to fame was that once, when working for British Waterways, he accidently drained the Chesterfield Canal by pulling out the plug. Anyway, Steve convinced me that the little tablets he had with him were some sort of "upper" that would help me to carry on without sleep all night. Naturally, I took one only to discover they had exactly the opposite effect and I fell asleep in my cabin before even going to the bar and slept soundly until being woken up in Holland. Of course, there were many incidents involving alcohol on all the trips. In Rotterdam, I stood and watched as one of my mates "Iggy" swayed drunkenly next to the support column for the tram wires; fortunately, he swayed away from the tram as it passed, a sway in the opposite direction would have trapped him between tram and post and crushed him. In Hamburg, another of our party was minded to visit a "lady of the night" at the Eros centre. We later located him at the hotel having retuned from the Accident and Emergency department of the local hospital with a few cuts and bruises. His version is that he got cold feet and changed his mind about seeing the lady but in his dash to get away, he was knocked down by a passing car. At least that was his story.

Paris was also a good trip but did not involve night travel, instead we used the cross-channel ferry from Folkestone. On the way home, about 10 of us stayed too long drinking in a bar and in our rush to get the train home, we jumped on to the "Golden Arrow" at Gare du Nord only to find that our reservations were for a slightly later relief train. It was too late by then to get off, but we knew that that both trains would stop at Amiens to change engines. We calmly got off there and sauntered down the platform to greet the rest of the party as their train drew in. Jack, the trip organiser was not amused. Apparently, he had almost missed the train in Paris whilst searching frantically for us around the station.

The most outrageous trip of all was, I think, one of the last and it involved Brian Street. Brian worked in the drawing office but at weekends he would volunteer as catering staff on one of the many train excursions that were run in those days. Brian had good connections with the catering department, but he was also, we all thought, rather inclined to exaggerate what he could achieve. The restaurant car that used to form part of the boat train to Harwich had long since been withdrawn to be replaced by a small buffet, but Brian promised us that he would arrange something special for us. When the train pulled into Nottingham that Friday evening, it had attached to it a full restaurant car with kitchen, chef and waiters. The car was reserved for our sole use, and we were served a full 4 course meal. Of course, we paid for the meal but the cost of putting on the restaurant car would have far outweighed the money that we paid. The coach had had to be sent to Harwich the previous day from Stratford in East London then attached to the morning train from Harwich to Manchester in order to be with us that night on the return working. On our return journey on the Monday morning, we had the restaurant again, serving us full breakfast. Unfortunately, the excesses of the previous night meant not all of us could do it justice and "Iggy" actually fell asleep in his "full English".

There are more stories that could be recounted, some associated with the diving team which I might mention later. However, later in my career, I began to see the downside of drinking with good men losing their jobs and mistakes being made so my view of the drinking culture changed as I got older and, hopefully, wiser. By the time the rules came in to stop it all, I have to say that I was ready for them.

Chapter 9

Lady Elton's Personal Engineer

There was a period when I worked in the bridge section at Nottingham that barely a day went by without someone shouting across the office, "Brian, her ladyship is on the 'phone for you," and for a while the "wags" in the office referred to me as Lady Elton's personal engineer. The reason for this was Bridge 92 on the Midland Main Line, at Sutton Bonnington between Trent Junction and Loughborough.

Lady Elton was the owner of Sutton Bonnington Hall which stood just on the western side of the railway in the village of the same name. The hall has been owned by the Paget family since 1825. Sir Earnest Paget, the 1st Baronet Paget of Sutton Bonnington, was chairman of the Midland Railway from 1890 to 1911, and his son, Sir Cecil Paget, was a locomotive engineer and railway administrator for the Midland Railway and was once the manager of Derby locomotive works. Lady Elton is herself a member of the Paget family. She married Sir Rodney Elton, 2nd Baron Elton, in 1958. Sir Rodney was a minister in Margaret Thatcher's government. They were divorced in 1979 but Lady Elton remained as owner of the Paget family home so when I first met her, she was very much the sole person in charge at the hall.

At one time the Paget family had their own platform on the estate with a right to have trains stopped at their convenience and of course there were very close ties with the Midland Railway dating back to Sir Earnest Paget. Lady Elton was very aware of this and frequently reminded me of the fact. The Sutton Bonnington Estate extends on each side the railway and they were connected by accommodation bridge number 92 which spanned the 4 tracks of the Midland Main Line. In 1980, it was a single span wrought iron bridge with lattice girder main beams that also formed the parapets. This type of bridge is commonly known as a "through-deck" bridge. It was rather a handsome structure and was adorned with the crest of the Midland

Railway. Unfortunately, it was also badly corroded and deemed to be in need of reconstruction.

At the time, British Rail, along with other nationalised industries and public bodies were required to pay much less notice of the sensibilities of their neighbours than would now be the case and were inclined to act first and talk later. It could be argued that the pendulum has now swung too far in favour of local inhabitants to the detriment of our ability to build major national civil engineering projects. Anyway, in the case of Bridge 92, the London Midland Region Chief Civil Engineer designed a new welded steel bridge to replace the existing structure. As far as I am aware, this was done without having any contact with Lady Elton. A contract was awarded for the manufacture of the structure and the detailed drawings were passed to us in the bridge section at Nottingham to carry out the reconstruction using our own in-house team. I was put in charge of the project along with our major works supervisor, "Chalky" White. There was no road access to the bridge so it was determined that all major components would be delivered by rail and that rail cranes would be used in the construction work. However, we did need access over the estate land for concrete delivery, staff, and small materials. We also needed to agree a period of closure of the bridge for work to be carried out, so I made the initial approaches to the estate and they were handled by the land agent on behalf of Lady Elton.

The land agent was a pleasant man who understood our needs but warned us straight away that we would have a tough task when dealing with her ladyship. My initial letter had made her aware of what we intended to do, and the answer was an immediate no, "you will not touch the bridge and you will not come onto my land". Her ire was particularly strong, I think, because we had not approached her as a member of an aristocratic family with close links to the Midland Railway. I am sure that this was also allied to a certain contempt of nationalised industries. This of course put me in a difficult position. I was responsible for getting the job done; the steelwork was already being manufactured so there was no way that I could change the design to meet her Ladyship's objections and a possession date for the reconstruction had been set. We continued to talk to the land agent without much success and he told us she was really digging her heels in and investigating all the old documents relating to the original construction of the railway through the estate to find ways of stopping us. It looked as if we might have to go along the legal route to try to force her to let us do the work but even that might not be successful and would inevitably drag on while we continued to look after a bridge over the main line that was in a rather poor condition. I also had a niggling doubt about the validity of our case because if truth be told, it would have been possible to repair the existing structure and restore it to its original condition and appearance had our CCE not decided to go along the route of full reconstruction and if Lady Elton hired a structural engineer who could demonstrate this, then they might well win the day.

Eventually, the land agent got Lady Elton to agree to a site meeting to see if we could come to some agreement. I have to confess that I did not approach the meeting with any great expectations of success. There were to be 4 of us at the meeting, myself, and a surveyor from the property department on behalf of British Rail, the land agent and Lady Elton. The agent met us at the gate and took us over to the bridge. We then waited for her ladyship to arrive, which she did, late, mounted on a beautiful and very large and powerful (to my eyes) horse. Furthermore, she was dressed in fill riding gear of dark blue with riding hat and blue net veil. The expression "dressed to kill" came immediately to mind. Unfortunately, our man from the property department, who was there as an expert on the legal situation regarding rights of access, etc., was very afraid of the horse and each time it moved, he took a step backwards until he was so far away, any communication with him was by shouting. I explained why we had to replace the bridge, but Lady Elton was adamant that whilst she understood the need repair works, the appearance of the new bridge was completely unacceptable to her, and we must retain the lattice girders and Midland Railway crests.

We looked more closely at the bridge, Lady Elton from her vantage point on the horse and me from the ground and after some fruitless discussions I asked if the bridge really needed to be so wide. It transpired that this bridge was not the main route used by farm vehicles and that they could compromise on width if I had a suitable suggestion. The land agent was very helpful in this respect as we began to hammer out a workable compromise. The proposal that we came up with was that the bridge would be rebuilt, as designed but that the two lattice girders would be fully refurbished and brought back to site where they would be placed in permanent positions on the inside of the new main girders. Thus, the appearance of the bridge seen by people crossing it would be maintained as the Midland Railway bridge even though all the loading on the bridge would be taken by the new structure. We agreed that I would put the proposals in writing and the meeting broke up without her ladyship even dismounting from her horse. In the end, she accepted the proposal, and our relationship became more cordial although it remained one very much of master and servant. I leave you to guess who was who. I got on well with the land agent, but he was so fearful of getting in bad books with her ladyship that he never let us take any liberties during the course of the works. He did tell me one anecdote. Apparently, Lady Elton's relationship with the villagers of Sutton Bonnington became so bad at one time that the local vicar had to pluck up courage and ask her to stop treating them as her personal serfs!

Having made promises to return the girders to site in good condition, I now had to deliver the scheme in its entirety. Clearly, the budget would be stretched but, in those days, there was no asking for extra money. All we could do was to contain the extra costs within the original budget as far as possible, put a little extra money to the job that had been saved on other projects and hope that the overspend would be contained within acceptable limits. Fortunately, money had been included for the possibility of having to build an access road to the bridge and I found ways of avoiding this cost which helped considerably. It now became a more straightforward engineering project in essentially three phases, all of which were interesting and not without difficulty.

Phase one was the reconstruction itself and the main problem was that both the old bridge and the new one were single span bridges that crossed all 4 lines. The maximum period that all 4 lines could be closed was for 3½ hours early on a Sunday morning and even this required some retiming of the returning empty newspaper trains. Three and a half hours is not long to take down a bridge, prepare the tops of the abutments and place a new one. Time being of the essence, the following plan was agreed. Firstly, the Up and Down Goods lines would be blocked from early Saturday afternoon and all traffic diverted over the two passenger lines. The goods lines were paired together on the eastern side of the line. The steelwork was then brought to site by train and unloaded on the goods lines. The main girders were placed into position on stillages and connected by the cross-girders, all of which were bolted together using High Strength Friction Grip bolts. Fibre cement shutter boards were then placed between the cross-girders to provide permanent shuttering for the in-situ concrete deck that was to be cast later. All this was done before the block of all 4 lines became due.

To do the actual reconstruction, we used both Steam Breakdown Cranes from Toton Traction Maintenance depot. Before the main possession, the 75-ton crane was attached to the new bridge and a test lift was carried out to see that the bridge rose evenly on the chains. At the same time the 36-ton crane was positioned ready to lift down the old bridge in one piece. This crane was Charlie Prince's old crane that had moved from Colwick to Toton when the former shed was closed. As soon as possession was taken, the 36-ton crane was attached to the old bridge with chains and lifted into the air whist the new bridge was also lifted up ready to be placed into position. It was an amazing sight to see 2 bridges suspended in the air over the railway at the same time. Fortunately, preparation work to the abutments was minimal so well within the 4-hour block, we were able to land the new bridge into position and place the old bridge onto the stillages on the goods lines for dismantling. The rush was then over, with the permanent shuttering in position on the new deck, we were able to work above the lines without interfering with traffic and the old bridge could be cut up and loaded onto wagons while traffic passed on the main lines. By Sunday afternoon, this phase of the work was complete and all lines had been re-opened. Care had been taken to load the main girders onto long bogie bolster wagons in a way that they could be forwarded on in normal traffic.

Phase 2 and 3 continued concurrently. Phase 2 on site and Phase 3 elsewhere. On site the one remaining major job was to place the in-situ concrete deck which had to be cast over the full length of the bridge. There was no way that the concrete could be delivered up to the bridge by lorry, Lady Elton would not give permission and even if she had, the cast of constructing a road across the steep field would have been prohibited. The only way that we were even allowed on site was by making the absolute minimum of disturbance. Because of this, we elected to pump the concrete from the nearest public road in a 300m long pipeline. This was close to the maximum length that concrete could be pumped, especially uphill. No mechanical plant was allowed on the field, so each section of steel pipe was carried by hand and all joints were wrapped in polythene sheeting to prevent seeping concrete from contaminating the grass. The pumping was successfully carried out in a day and when it was complete, each pipe was taken away and cleaned off-site. At the end of the day men could be seen sweeping the grass and picking up any small spillage before her ladyship turned up on her horse for a round of inspection.

Meanwhile, away from the bridge, Phase 3, the refurbishment and return of the old main girders was in progress. Firstly, we attached labels to the wagons and put them in traffic to Woodburn workshops at Sheffield. At the time this work was being carried out, wagonload traffic was on the decline, but it was still possible to move loads in this way. Once they arrived at Woodburn, our good friends and colleagues of the Sheffield steel workshops fully refurbished the girders, an operation requiring extensive hot riveting. I was determined that the girders should be returned to her ladyship in "mint" condition, so once the refurbishment had been completed the girders were again loaded on to rail wagons and this time despatched to Doncaster Belmont Yard from where they were moved to the works of Jesse Tighe on the outskirts of the town to be blast cleaned and painted to full BR specification and in a colour that Lady Elton found acceptable. Once this was done the girders were again loaded back onto rail wagons and moved in traffic to Toton yard from where, one Saturday night, they were formed into a ballast train with the Toton Steam Crane and taken to Sutton Bonnington to be placed in the final position on the bridge. That completed the bridge works and I was more-or-less released from my position as Lady Elton's personal engineer. However, for a few more years, whenever Anne Elton had a complaint to make about the activities of railwaymen near her property or some issue with the lineside fencing, it was my phone that always rang first.

Chapter 10

A move into Management

I have previously mentioned that most railwaymen at the time relied on overtime payments to maintain a reasonable standard of living, but these were only paid to the "workers". Once one entered the management pay structure, there were no overtime payments made and certainly no large bonuses of the sought talked about today. The only additional remuneration that engineers could get was a small allowance for being on-call. Accordingly, many technical staff and supervisors didn't even consider applying for management jobs, recognising that it would take a further 3 promotions before their take-home pay reached the money they were already getting when overtime was included. In my particular case, I was working 3 or 4 Sundays every month because if not required for weekend bridge works, I was working as a diver, inspecting bridges. When it was suggested, therefore, that I might be interested in the job of Works Production Assistant – which was graded MS1, the lowest management grade – I had some thinking to do.

On the one hand, I was very attracted to the job. The previous incumbent had been Fred Spreadbury, the Chief Works Inspector who I respected enormously, and the job of "chief" had been regarded as a very senior position in engineering departments. Admittedly, the job no longer carried the kudos of the old "Chief Works Inspector", but it still had all the same responsibilities and duties for all the outside staff plus control of the planning and "works study" staff. On the other hand, taking the job would result in a very substantial loss of pay. Fortunately, my work as a diver was recognised as being valuable so it was agreed that any diving that I did at weekends would receive a special remuneration separately from my role as Production assistant. It still meant that I had a small reduction in salary overall, but it was manageable, and I was in a much better financial position from many others who made the jump to management but never really recovered financially.

The job itself was pretty full-on and again brought me into more conflict with "spanner man". Operating through local works supervisors, I had to ensure that all the work carried out by our own staff was effective and efficient. I have to say that the quality of the supervisors that I had at my disposal could best be described as variable and covered a broad range of abilities and attitudes. I had depots at Kettering, Leicester, Melton Mowbray, Derby, Burton-on-Trent, Sandiacre and Nottingham as well as a steel gang, also located in Sandiacre and a workshop at Derby Friargate. Each depot had a works supervisor with up to forty artisan staff and other assistants such as timekeepers and there were also some specialists such as the plumbing supervisor. During the course of the job, some of the depots were closed so that in the end we had a smaller number of larger units. Some of the best supervisors were excellent and I would single out the team at Kettering as being very good whilst most of the others were competent but there were those that had drinking problems and at least one that, it transpired, who was "on the fiddle". The latter was discovered by a particularly vigilant member of my works study team who tended to act as my assistant in many ways.

He spotted that one of the supervisors was getting a large number of "callouts" on a Saturday morning from various workshops in the Derby area as well as from the Derby School of Transport. When we investigated it, we discovered that most (if not all) of the callouts were for spurious reasons. Getting proof of this would have required involving the railway police and would, if proved, meant the immediate dismissal of several otherwise good staff in both our department and others. The matter was therefore dealt with by giving out serious "bollockings" and the demand that proof would be required before any more call-out payments would be made. On the other hand, I was ready to support to the hilt any of my staff and supervisors against accusations from others including, of course, "spanner man".

One of these disputes arose over the precautions taken for the control of Legionnaires disease. There had been a spate of deaths from this around the country and instructions were issued for the treatment of water supplies and other precautions to be taken where showers were in use because the principal spread came via fine water droplets. I appointed Leo Seymore, our services supervisor to undertake the work and he set about it with gusto. There were showers in Toton Loco and, I think, in the Wagon Works at Long Eaton and at Derby Loco. But the biggest risk was felt to be at the Derby School of Transport which was a residential school and therefore the students were using the showers daily. Leo arranged for all the necessary water treatment and shower head cleaning but, of course, it could not be carried out when students were present, so it became a regular weekend job for Leo with Saturday morning being taken up with carrying out water treatment and a return visit on the Sunday to flush out the system. You would think that the company would be pleased that one of their staff was prepared to give up his weekends in this way but that is not how it was seen by "spanner man" and others. Most supervisors got regular weekend work on possession jobs but there had, up until then, been very little requirement for a Services Supervisor at weekends so Leo lost out on the overtime. The allegation was that Leo was exaggerating the Legionnaires disease work for his own benefit. Since Leo was the only one who had bothered to find out exactly what was required there may have been some truth in this allegation, but I was having none of it and defended my staff in a series of unholy rows with John. Of course, by then John and I were on a par in terms of our responsibilities, another thing that annoyed him.

The main tools that I had for carrying out my job were regular site visits, both to worksites and depots and the weekly "Farming-out" meetings. The latter were when I met with each supervisor in turn to discuss the progress of work over the previous week and planning the work for the following week. They had been part of the system since long before I had joined the railway and in district days it was possible for the Chief to visit each depot in turn on a weekly basis and as well as the farming out, he would do his weekly inspection of the work of the depot. The size of the new divisions, however, made this impracticable and I know that in some parts of the country, the procedure was abandoned. I felt that the weekly meeting was essential to maintain control over my supervisory staff, particularly those who disliked any form of control of their authority. So, I arranged for one day a week to be farming out day when each supervisor in turn would visit me in the "farming-out room", which was a small side office that I had managed to get equipped with a computer. Starting early in the morning and going on till mid-afternoon and sometimes later, I met with each supervisor, and we did all the progress reporting and forward planning of their work which I recorded on a spreadsheet that was a sort of forerunner of "Excel" but much less user friendly. The meetings could be fraught when things were not going well and sometimes I felt that I had made a rod for my own back, but they did enable me to hold on to a degree of control.

I got most wound up with the steelwork supervisor who was nicknamed "anvil". He never seemed to be able to get the work done that I wanted or to provide me with the information I needed. It was also true that his "spot" was in the afternoon when I had already had a number of issues to deal with so one particular day, I lost my cool completely and kicked a metal waste-paper bin at him which caught him squarely in the face but fortunate without the force necessary to cause injury. Despite all that, at the end of the day, I would finish up with a computerised list of work completed, work planned, and progress made including number of man-days spent against the estimated number required and I was proud of this since very little other use was being made of the new-fangled computers at the time.

The information that I gathered at "farming out" armed me for my attendance at the monthly progress meeting where the Assistant Civil Engineer (Works) would go through all the larger jobs that were in being carried out with particular reference to the annual budget which could not be exceeded under any circumstances. Others at the meeting were the chief clerk, the head of the bridge section and, of course, "spanner man". I remember one occasion when John berated me for erecting a new fence at Bingham Station which did not match the fencing on the opposite platform. This really annoyed me because instead of paying out good money (as usual, there was no money for station repairs in the budget) for new material, we had used some fencing left over from the new station at Alfreton and the lads had done a good job. It was also of a simpler form, so I reckoned that I had saved over 50% of the overall cost. Neither of us would back down until the chairman closed the argument. The good thing was that the print-offs from my computer contained accurate progress information which very much aided the purpose of the meeting and that I could use to refute some of the allegations about the effectiveness of our own staff. There was a view held by some managers that we would be better not having our own staff and letting all the work out to contractors. I vehemently disagreed with that approach and always defended our staff but sometimes it felt like I was fighting a losing battle. Later, when privatisation came along, all the work was put out to contractors, and I firmly believe that was to the overall detriment of the railway.

The site visits were when I could get to know the men better as well as the supervisors, but I already knew most of them by name because I had met up with them whilst doing my bridge and building work. We had many interesting conversations and some successes with the projects that we undertook but there were inevitably conflicts that sometimes got escalated into full blown disciplinary procedures. I learnt a lot during this process both about the practicalities of railway engineering and about man management as well as dealing with the problems that working life throws at us all. As any manager will tell you, there is always at least one member of staff whose sole purpose in life is to cause trouble.

My nemesis for several years was Ken Ward, a plumber at the Leicester depot. General maintenance work was the main purpose of my works organisation but there was never enough money in the kitty for this so some larger jobs were undertaken, and we could also make a play for "New Works" projects. I saw these as opportunities to demonstrate the effectiveness of our team and also to provide us with more interesting jobs. One such job was the refurbishment of the Gentlemen's Toilets at Leicester Beal Street Loco. Ken Ward was given the job and duly installed a bank of urinals that were too high for even the tallest engine driver to reach. Some hilarity ensued with attempts by the younger members of staff to reach the height from a distance and others bringing in off-cuts of sleepers to stand on. Of course, "spanner man" got hold of it and used it as another reason to "rubbish" my department. The drawings supplied from John's section were OK, but they did not specify the height to which the urinals were to be set. Any reasonable man would have used his own good judgement or asked for advice. Not Ken, he deliberately installed them too high, and I could not defend him. Eventually he was put on a "Form 1", which is a disciplinary procedure that resulted in a reprimand but which set in motion a chain of events that caused some embarrassment for me.

Sometime later, we had to install a new water main to the Carriage and Wagon depot at Wigston to replace an old iron main that was the cause of several "bursts". As far as I and the Leicester supervisor was concerned this was a purely domestic issue on the railway main and we had no need to involve the Severn-Trent water authority. It has to be said that we were generally quite cavalier with issues like this, relying on the sweeping powers included in a number of railway acts to enable us to ride rough-shod over local authorities and others. Of course, Ken thought differently and having installed the new main in its trench he refused to back-fill it until the water authority had carried out an inspected. The matter escalated up to me and I was determined this time to teach Ken a lesson and I issued him with another "Form 1".

At his disciplinary hearing, Ken alienated his own Union representative by arguing with him and accusing him of being a management "lackey", so I had no problem in finding him guilty and deducting him a half-day pay and adding another black mark on his record. However, Ken wasn't finished, and he sent his story to the "Leicester Mercury" who published it in full and included my name, suggesting that the work had been ordered to by-pass a water meter which, of course, was untrue. He also took his claim to the small claims court and won his half-day payment back. I was furious but the railway legal department knew that I was on dodgy ground and ordered that no further action was to be taken. The matter did no credit to anyone, least of all me and I resolved to be surer of my ground when dealing with any similar matters in the future.

My office staff consisted of 4 or 5 works study staff whose main job was to quantify the details of each job and use the information to establish how long it should take. Targets were set for the number of "allowed minutes" for each job and the actual progress was monitored. The staff were then paid a bonus that was calculated from the progress achieved. It was a rather cumbersome system, and some doubted its effectiveness. It was certainly alleged that more bonus was earned with the pencil when filling in the worksheet than was actual earned with the tools. Set against that, it was a useful tool for planning and monitoring the work. The best workmen could also earn a good bonus by working hard but there were constant conflicts with those who wished to try to claim for work not done.

Since much of the scheduling had to be subjective, it was also a useful tool for me to reward staff for initiative. In a nationalised industry, managers can't hand out random rewards to staff but if I saw the need, I could arrange for an 80 performance to be allocated which would ensure a good bonus payment. Occasionally, there would be a major issue raised and one of these arouse at Derby where we were to repaint a whole office and welfare block at Etches Park depot. The work study assessor, Gordon Issott had produced a schedule but the painting team, supported by their foreman said that the time allowed was grossly insufficient. To further their claim, they made the mistake of eliciting the support of "spanner man" in his capacity of head of the building section. John was normally complaining that the men did not do enough work so it occurred to me that his intervention was probably a way of stirring up trouble, so I decided to make an issue of it. I took Gordon with me to Derby and in the company of the painting foreman I personally measured each wall, door window, and skirting board then applied the formulae to produce a schedule. I was able to prove that Gordon had been more than generous in his assessment gave him a mild reprimand and told the painters to get on with it. They had no choice but to comply. Gordon was a member of the diving team and a good mate of mine, but his laid-back attitude made some people think that he was less than efficient at his work. I had to make sure that our friendship did not affect my judgement of him as an assessor but in fact, I found him to be good at his job although inclined to find ways of getting off early in the afternoon.

At the other end of the spectrum was Mike Alexander. He was the son of the Permanent Way Renewals Manager and a real live wire, determined to impress and with what might be described as a strong sense of "right a wrong". He was allocated Leicester depot and as well as his Works Study duties he took it upon himself to be my eyes and ears looking for any wrongdoing. Leicester was not the best depot and we were trying to pull it round after the departure of the previous supervisor. Sometimes, I had to rein him in a little but when he came to me with proof of heavy drinking at work by a bricklaying team, I had to take action. The two men concerned were repairing a bridge near to Melton Mowbray Station and Mike reported that he had twice found empty beer cans on site and that when he visited at lunchtime, the men were nowhere to be found. He and I went together one day at lunchtime, and I was intent on challenging them when they returned to the site. By 2pm they had still not returned to the bridge, so we set out to look for them. Mike had done his homework and led me straight to a pub near the station where they sat drinking with the landlord and a couple of locals. Despite some aggression from the landlord, I asked them to leave and escorted them back to site to collect their tools before taking them back to the depot and suspending them from duty. I reflected that in the old days, I would have told them to walk home, minus any railway clothing they had on. They resigned before they were sacked because the union rep had told them he would not give them any support. The unions were as keen to stamp out drinking at work as the management.

I felt that I had made progress while in the post and could point to a lot of good jobs that we carried out. I was also particularly proud of the apprenticeship scheme that I re-vitalised. I had two excellent apprentices at Derby. One, a carpenter, eventually left the company to work for himself and the other, Neil, went on to be a senior manager with Railtrack and later Network Rail. He was a good bricklayer but had a temper and did tend to get himself in hot water. Sometime after he had moved away from the tools and into a management position, he joined the Furlong House staff on a trip to Southwell races. Next to the racecourse is the small country station of Rolleston and on race days, extra trains stopped there. At the end of the platform there is a level crossing manned by a crossing keeper. For some reason, on the return journey and having consumed a few pints, Neil got into an argument with the crossing keeper which escalated into a full-blown fight. Neil was a big bloke, so the crossing-keeper locked himself in his cabin and rang for the Railway Police. Fearing that we could all lose our jobs over the incident but determined not to leave Neil to face the music alone, we hustled him on to the first available train before the police arrived. Unfortunately, that was going away from Nottingham and took us all to Newark. We then caught a train back which stopped at Rolleston again. We all slumped down in our seats so as not to be recognised but we needn't have worried because we glimpsed the crossing keeper being forcibly pushed into the police car. We found out later that he had been so aggressive with the police that they arrested him, and we heard no more of the incident.

I have mentioned the Italian workers who came over after WW2 to work in the brickyards at Peterborough, but many others came to work on the railway. In Nottingham we had a good number still there in the 1980s, including several DeRosas. Mr DeRosa senior was an immigrant and he worked in the depot in Nottingham. In turn several of his boys joined the railway in various positions including Andrea who worked in the clerical office at Furlong House. His younger brother, Pino joined us as a Labourer in Lenton Yard, mainly being employed in stripping down panels of track that had been brought back from relaying sites. Andrea was determined that Pino should get a good career and pestered me to arrange for him to be transferred to the works department and be given an apprenticeship to be a bricklayer. I resisted at first because theoretically, Pino was already too old to start as an apprentice but eventually I succumbed and he got his transfer. It was one of the best things that I ever did. Pino was an excellent worker and a quick learner. Eventually he came into the technical office and joined the diving team. Much later, when privatisation forced me to leave British Rail, Pino and I, along with Andy Jones, formed a partnership and set up our own business. Although I have now retired, it is still flourishing under Pino's guidance and in 2020, it celebrated a 25th birthday, furthermore several more De Rosas have joined the firm over the years to continue the family connection.

One of the "perks" of my job as Works Production Assistant was to take part in the annual works inspections. The format was that supervisors would each submit proposals for work to be carried out based upon bridge examiners reports and their own judgement. I did a certain amount of guidance and vetting in the process and when this was complete, all the structures proposed for work had to be inspected by the Divisional Engineer, supported by his assistant managers and the supervisor concerned. Many of the inspections were carried out by road but when rail traffic conditions permitted, we used the inspection saloon. This was a rail vehicle that was towed or propelled by a loco and had a large window at the back from which the overall condition of the line could be inspected as the train passed by and which would stop at each structure in turn. When it stopped, we all alighted on to the track and the supervisor would indicate the work that he proposed, and the engineer would give the work a priority and allocate a sum of money for the work. All this would be recorded by the programme clerk who accompanied the party and then we would get back aboard and move on to the next structure.

In general, only Priority 1 jobs were sure to get into the budget for the following year along with some P2 items but P3 and below would have to wait for several years until money became available. It was a good system because it enabled a slowly deteriorating structure to be monitored until they could be repaired. It was expected that an item would enter the system with a low priority and gradually move up the tree. If a supervisor submitted one to be seen for the first time that was already in poor condition he would be closely questioned as to why it had only just been brought to the attention of the engineer. It was easy to use the saloon on freight lines and on lines where there were more than 2 tracks but on busy 2-track sections we would be herded in and out of the saloon with the words "2 minutes only here or we will delay the express" and we also had to wait for some time in sidings before we could proceed. Because of this, the 2-track main line from Leicester to Kettering was normally done by road but on other lines we could have a very pleasant and leisurely day. Favourites were the Leicester-Burton Line which is freight only and at the time carried mainly coal traffic and the line from Manton Junction to Kettering, which was little used and passed through several tunnels and over the magnificent Harringworth Viaduct.

My first trips were when Malcome Harbottle was the engineer and during the time before the drinking ban came in. We would set off around 8am and have coffee and biscuits while we travelled to the first site and carry out inspections until lunch when we would stop in a siding and sit down to a cooked meal prepared on the train. However, once 11am arrived we were offered beer or other alcoholic drinks when travelling between sites and wine was served with lunch. Inspections carried on all afternoon and sometimes into the early evening. A day on the Inspection Saloon was also a way in which the engineer could impress his colleagues so we would also have guests on the train, and we had to make sure they stayed safe. Alcohol and a live railway do not really mix! The most interesting saloon trip I had was over the old Clowne Branch. This line in North Derbyshire/Nottinghamshire joined Barrow Hill to the Nottingham-Mansfield line but by the time we traversed it there was little, if any, traffic using the line and it was heavily overgrown. Fortunately, we were travelling engine first and the driver attacked the vegetation with gusto, and we left behind us a trail of broken branches and other vegetation.

By 1988, I felt that it was high time that I moved on. My relationship with John Thompson had not improved and it began to feel that my working life was one series of battles. I began to scan the vacancy lists but with children still at school and a well-established social life in Bingham, I was not keen to move house. I did feel, however, that by driving to Grantham and catching a train on the East Coast Main Line to London or York, I might be able to commute. It took some time before I spotted a job that I thought would really suit me and the last few months was not a pleasant time. I had about 50 years working on the railway, and I can honestly say that for all but 2 periods, each of about 6 months, I was very happy at work. This was one of those two periods. Looking on the bright side, this means that for 95% of my time I had a happy and satisfying career which is better than most people can say.

Chapter 11

Tunnel Work and the 100 MPH Tunnel Train

Tunnels deserve a full chapter in the book, and it is one subject that spans the whole of my railway career. Peterborough District had 3 tunnels to look after, all close to Grantham. Stoke Tunnel was on the ECML just South of Grantham while Peascliffe was also on the main line just North of Grantham and Gonerby was on the Grantham to Nottingham line. The first time I went into one was at Stoke tunnel on a Sunday morning when I was sent there to measure up the quantity of repairs carried out by contractors who were employed to cut out and replace defective brickwork. To a 17-year-old, it was a bit like entering the fires of hell. I walked in from daylight carrying my "Tilley" lamp into the semi-darkness of a tunnel filled with fumes from the generators and compressors in use as well drifting fumes from the ballast train loco, idling just outside the tunnel portal, all accompanied by a cacophony of noise from compressed air "shoulder guns". There was some pressure on me as well because my measurements would not only determine the payments made to the contractor but also to the payments that the men would receive. I found out later that a predecessor had been sacked for accepting a packet of cigarettes from the contactor's foreman that contain £50.00 in notes! If anything, the second tunnel visit, this time to Peascliffe, was even more traumatic because it was a mid-week site visit carried out while normal traffic was passing through the tunnel at 90 MPH. This was normal practice until the end of British Rail, and we all knew that it was important to know exactly where each "refuge" was so that we could find it quickly when a train entered the tunnel. Even in the refuge, the noise and blast of air when a train passed was frightening and later on, bars to hold on to were fitted in many refuges because of the "pull" when trains travelled through at faster and faster speeds.

Wewlyn South Tunnel on the East Coast Main Line

When I moved on to King's Cross, we had many more tunnels to look after; 9 on the ECML and 2 long ones on the Hertford Loop as well as several for suburban services including Hotel Curve, which passed under King's Cross Station itself. Potters Bar Tunnel had been widened some years before, so the original bore now carries the down lines and the new tunnel is for the up lines. The 2 tunnels are connected with cross-passages so standing in one bore, the sound of a train passing through the other can be most disconcerting. As usual at weekends I was working for the P-Way department and for 5 or 6 consecutive Sundays, I was in Potters Bar Tunnel on a relaying job controlling the depth of new ballast being spread by "Traxcavator" before the track panels were laid.

It was mid-summer and very hot. As the dry ballast was tipped from wagons and then spread by the machines, the dust completely enveloped the whole of the workforce and when we finally did get out of the tunnel at the end of our shift my mouth, nose, and every crevice of my body was filled with it.

In those days, the traxcavators had no cabs and one driver called "sailor" stood up as he left the tunnel and while still on the move dropped his trousers and pants and wafted his manhood in the breeze! Hotel curve tunnel was very tight, and access was only allowed in possession, it was also on a steep gradient. Down trains from Moorgate came out into the open alongside King's Cross suburban station on Platform 16 whist up trains ran though York Way tunnel, which was on the east side of the station. As well as passenger trains, some cross-London freight used the line although that was already dwindling when I arrived. I never had occasion to go down, but a colleague had the job of testing the rails for cracks, a task which involve oiling the top surface of the rail. Unfortunately, he forgot to clean the oil off before he left and the first train to use the line afterwards sipped to a stop on the gradient causing severe disruption to the commuter service.

I had little work on tunnels while I was at Leicester although we had several tunnels on the patch, including a very long one at Corby. Generally speaking, these were looked after by the Bridge Section in Nottingham. There were, of course, occasional tunnel visits but only when I moved to Nottingham did I come into close contact with tunnel works again. I have no idea exactly how many tunnels we had on Nottingham Division but there were plenty, especially on the Derby-Leeds line, the Matlock branch and the Manton to Kettering line; so, although Alan Hughes was the tunnel expert in the bridge section, all of us got to work on them at some time and I was also heavily involved when I was Works Production Assistant. Most of the maintenance work was basic brickwork repairs and required very little technical input. If the bricks themselves were loose or badly eroded (spalled is the correct term) then the brickwork would be cut out in panels and replaced. If the mortar in the joints was soft or loose, then mechanical re-pointing would be employed. The key thing in all this work was to ensure that both men and materials could be quickly got into the tunnel and a safe platform provided to work from for the full height of the tunnel. It was also necessary to ensure that both men and supervisors knew exactly what they were doing because it was a skilled job. For that reason, the teams were always made up of experienced bricklayers and assistants and whenever possible, apprentices were put with them to learn their trade.

When I started in Nottingham, the tunnel train included a wagon known as the "Queen Mary" which comprised a timber platform built on a 60-foot long "Sturgeon" wagon. Unfortunately, this wagon was eventually condemned so we had to make up a new train, a task that I left to "Chalky" White, the major works supervisor. He did an excellent job and we finished up with a train that included 2 wagons fitted with permanent scaffold, several open materials wagons and a long wagon onto which was fitted a mess room and secure plant store.

It was during this time that I learned from the skilled bricklayers exactly how to repair tunnels. This was to stand me in good stead later when I moved on because I was able to provide training to new staff on how to do the work properly. The first rule that I was taught was that any brickwork repair had to be to a minimum depth of 225mm (2 rings of brickwork). Anything less was unacceptable as it would leave a single skin of new brickwork that could split away from the main body over years of wet and freezing conditions. This was a sensible rule, but it could mean very had work for those men "cutting out" because after the face ring, the second ring of brickwork might be perfectly sound and therefore very hard to break out. It was then that good supervision was required to ensure that the second ring really was removed!

After the first 2 rings were taken out, then a "block tie" would have to be cut out into the third ring so that the whole mass of new brickwork was secured to the body of the tunnel lining. Before any cutting out could be done, the existing brickwork had to be secured around the area to be repaired. This was done by bolting steel ribs to the tunnel lining and then inserting timber laggings between the ribs at top and bottom of the area to be cut out. The laggings also support the new brickwork as it is laid. The operation straightforward in the sides of the tunnel but towards the crown, the work becomes progressively more difficult until at the crown itself, the bricks are being laid horizontally on the laggings and "crowning-up" boards are also needed. It needs diagrams to explain all this adequately, but rest assured it is a job that requires skill, experience, and care if the finished product is to be both strong and tidy. The work also generated all sorts of semi-technical terms that had to be learned, my favourite of which was "course of pig" which is when the bricklayer doing the repairs has lost his way and fails to match in properly with the existing work.

Soon after I became Works Production Assistant, questions were being raised about the productivity on our tunnel repair work so, together with Chalky White, we set-to to improve matters. One of the first things that became apparent was that the exact areas to be done were not being accurately specified or recorded; I felt what was needed was to set amounts of work that I would be satisfied with, and which could be justified by Works Study Data and then find a good incentive to ensure that the work was always completed. Bonus payments did apply to the work, but they could all too easily get lost in late starts and the time taken to set up the train on site so something more direct was required and I opted for the good old "Job and Knock" principal. Basically, it means that when the work is done you can go home and will still be paid for the full shift. It has advantages all-round, particularly on possession work because it means that the job will be completed as quickly as possible, and the staff can get off home as soon as possible. However, care has to be taken because there must be enough work specified to ensure that it would normally take a full shift to do it and supervisors who themselves liked to knock of early had been known to set up short jobs and thus abuse the system.

To counter this I established a strict rule-of-thumb quantity that I would accept. A team of bricklayer and mate had to cut-out and replace a minimum number of square metres of brickwork in their shift, including all temporary works. The amount reflected the difficulty of the work and was set at 2 square metres for sidewalls, 1.5 in the lower parts of the arch (known as the haunches) and 1 in the crown. This could be a daunting target in hard brickwork but if the brickwork was soft, it could soon be achieved, and we could all pack up early. Of course, "spanner man" got in on the act and claimed that he had visited sites early in the morning to find everyone had already gone home but I stood my ground and productivity improved. Later, when this sort of work was taken over by contractors, they never got near this level of productivity.

Re-pointing of joints in tunnels was mainly done mechanically using a system known as "Aerocem". First on the shift was the man (almost always the same one) who used high pressure water jetting to wash the loose mortar out of the joints. This was a filthy job because all tunnels are covered in soot from steam engines, and everyone stayed clear until he finished. Then the "pots" were set up and the special mortar was mixed. The mortar then goes into the pots which are clamped shut and pressurised with compressed air. The air forces the mortar through a pipeline to the heavy "guns" which are used to fill the joints. It is dirty, noisy and very tiring work; what is more, although a good team with well-maintained equipment can produce hundreds of square metres of pointing in a shift, even a small blockage in the system can cause endless delays. Our Aerocem teams were excellent but in novice hands, the chance of getting any real productivity on site was nil.

The Nottingham division tunnel train – A forerunner of the famous
100MPH train
Photo – Brian Maddison

When I moved on to Peterborough, I found that although there were plenty of tunnels on the area, they had no tunnel train and were making do with setting up scaffolding in a random fashion for each job. Productivity suffered as a result, and I looked for a way of providing a new tunnel train at minimum cost. I tried at first to get hold of a pair of Freightliner wagons on which to build permanent scaffolds but there were none to be had. Freightliner vehicles were only used for transport of containers and therefore had only a skeleton body with no floor plates and using one of these would avoid the problem of build-up of rubbish under the scaffold. They were also very low wagons which would also have been useful.

Having been rebuffed in my efforts, I looked for an alternative. Each office had a book of wagon diagrams and leafing through it I found a "Bullion Wagon" that looked very like a freightliner vehicle. A quick check on the "TOPS" computer showed that there were 2 in a siding somewhere in the western region. I made enquiries and found that these had been built to carry containers of cash for the Royal Mint and banks and would be attached to passenger trains for long distance secure transport. At some stage the railway had lost the contract for the work, so they were now redundant but still serviceable. I managed to get them transferred to our department and then sent to Bounds Green Depot in London where we got them placed on a dock siding next to our Palace Gates Depot.

In the meantime, I had contacted the technical department of SGB scaffolding and I showed their representative our wagons and told him what I wanted. Between us we chose to use their "cuplock" system which meant that the scaffold could be extended on site without the use of scaffolders or special tools. SGB designed the scaffold and provided detailed drawings and supplied the components. We put it together with our own staff and fitted it to the wagon with the main attachment points being welded. The core of the scaffold ran most of the length of the wagon and consisted of a single platform level that was in "gauge" and so could stay in position when the train was on the move. This was fitted with stairs at one end. Once on site, this platform could be fitted with handrails and could be extended upwards and sideways by simply clipping on additional brackets and standards. To get to the crown, the very long brackets, once erected, could be propped from ground level to provide a stable platform. To form the working train, these platform wagons were attached to GUV wagons that were ex-parcel vehicles with end doors. These held all the materials and plant needed to repair tunnels and this could all be transferred through the end doors and up the stairs to the working platform without being unloaded to the ground. Beyond these, open wagons were used for compressors and generators. Fixed electrical routes and compressed air lines were installed so that tools and lights could be plugged into points on the platform. I was thrilled with the whole set-up and I sure it was a set-up that would greatly increase site safety and productivity. A review of risk factors was carried out and concerns became apparent about the possibility of young vandals climbing on the scaffold when it was in sidings, so we fitted large plywood barriers to make this more difficult as well as barriers across the staircase. Both wagons were fitted in the same way, and they could be used together on one train or could be used to make 2 trains.

Eventually both scaffold wagons were complete, and I arranged for a loco to move them from Palace Gates to Peterborough where they would be united with the GUV wagons and made ready for scheduled work in Stoke tunnel. I sat in the office at Peterborough, having checked that they were on the way and looked forward to seeing them pass by. I was chatting to the bridge assistant, Eric Jeyes, about the wagons and mention was made that the wagons were rated at 100 MPH. I suddenly realised that I had done nothing to change that, and they had just been picked up by a class 47 loco (top speed

95 MPH) and were heading along the ECML (speed limit 125 MPH). That meant there was nothing to stop it travelling at almost 100 MPH all the way from London to Peterborough and if it was signalled on the fast line, that is certainly what it would do. I had visions of scaffolding and plywood side boards being deposited all the way along the line. I anxiously watched out for its arrival and was very relieved to see it pass by with all parts intact.

After that, I took steps to down rate the speed to 60 MPH which I thought was quite fast enough for such a train. On site, it was a great success, we used it regularly up and down the area and one weekend, we really had the opportunity to test how good it was. A long weekend possession was arranged for some major relaying work between Peterborough and Doncaster with trains diverted via Lincoln. It was decided to take advantage of this to carry out large areas of brickwork repairs in both Stoke and Peascliffe tunnels. Two tunnel trains were prepared, both of which contained one of the scaffold wagons together with support vehicles and messing accommodation. We had continuous working all weekend with men coming from Doncaster, Peterborough, and Hitchin Depots to fill the shifts. We were able to renew large areas of brickwork in each tunnel and I felt that I had more than justified the cost of the train.

Sometime later, there was an incident at Peascliffe Tunnel which almost resulted in a major problem. I was on-call manager for the weekend and having visited several worksites at the south end of the patch, I decided to look in at Peascliffe Tunnel on my way home. Brickwork repairs had been scheduled to be carried out in a Saturday night/Sunday morning possession and the line was due to re-open at about 10am. I did not get there early enough to go in the tunnel and when I arrived, the train was just emerging from the south portal and was heading south back to Peterborough. To my horror, one of the standards that supported the handrailing on the wagon had been left up and as the train approached, I could see that it was very close to hitting the OHL wires as it passed underneath them. If the train ran out of the possession without being stopped, there was a real danger that it would strike the overhead wires when they dipped down for a bridge, and this could have resulted in miles of overhead line being fetched down. Both the supervisor and I were shouting as the train went by, but I doubt whether the driver had noticed us. Fortunately, a team of signal engineers working about half a mile south of us noticed us waving frantically and realising we wanted to stop the train, they stood where the driver could see them and held up both arms which is the correct hand signal to stop the train. The driver responded and the train came to a halt, but I was still concerned because I was afraid that the train might have gone into a section where the OHLE was "live". If that was the case anyone attempting to get on the scaffold to take down the offending post would be electrocuted. Fortunately, we were able to get it confirmed that this was not the case, and I went up myself and took down the post and arranged for the train to proceed.

We continued to use the tunnel train right up until 1995 and we made great play of its effectiveness in presentations that we made in the lead-up to Privatisation. During that period, we were encouraged to believe that if we could show that we had an efficient organisation, there was a good chance of it surviving into the private sector. Sadly, nothing we could have said or done would have been of any use. A "political" decision had already been made that the railways own works organisation would be disbanded in order to open up the industry to contractors. As a consequence, almost all my staff were made redundant, and the tunnel train was left to rust in the sidings. I was told that it would not be offered to the contractors as none of them would want the cost of hiring locomotives to move it to site.

I saw the effect of this a few months after privatisation. I had been hired through my new company to inspect work being carried out for Railtrack in Stoke tunnel. The contractor concerned had no previous experience of tunnel repairs, his bricklayers were hired in for the one shift and normally worked on building new houses, the plant and scaffolding were pushed into the tunnel on a small rail trolley and the breakers to remove the old brickwork were too lightweight to be effective. The result was that little or no progress was achieved, and this went on for week after week. Whilst the train was still available, I made several attempts to persuade Railtrack of its worth, even pointing out the safety issues being created but it was all to no avail.

In 1993, I was admitted as a Member to the Institution of Civil Engineers through the scheme for mature candidates. Aside from being an achievement in itself, it was particularly significant for several reasons. Firstly, as a chartered engineer and MICE, I was able to be permanently appointed to the post of Assistant Area Civil Engineer (Works). I was already carrying out the duties of the job, but the appointment could only be made to a Chartered Engineer. Secondly, I became authorised to carry out the inspection duties of a tunnel engineer, again, being a Chartered Engineer was a pre-requisite for this task at the time. Bridge and tunnel examinations were carried out by trained bridge examiners but for tunnel examinations, they had to be accompanied by a Chartered Engineer who had been formally approved for the work.

For my remaining time with BR, I did most of the tunnel exams on the Peterborough patch, so I got plenty of experience under my belt. Soon after privatisation took place, Railtrack appointed Civil Engineering Consultancies to undertake all of their bridge and tunnel examination. They soon realised that providing experienced and qualified tunnel engineers week after week on Saturday Night tunnel examinations was well beyond their capacity. Even if they had enough qualified staff, there was little appetite amongst their own staff to turn out at weekends. My company, Bridgeway started to receive many requests for help, and I was happy to oblige. As well as myself, I was able to contact three other former railway engineers, including my old boss, Alex Turner, who were qualified for the job. We set up Quality Assurance procedures to ensure that all 4 of us held Railtrack-approved certification and then proceeded to carry out hundreds of examinations over the years. The tunnels that we covered ranged from Whitehaven Tunnel, on the Cumbrian Coast in the north to a disused tunnel on the closed Somerset and Dorset Railway in the south. I often seemed to get the more remote tunnels such as the Settle and Carlisle line and also around Buxton in Derbyshire. The work was not unpleasant, involving arriving at site to be met by the contractor's examination team, many of whom were old friends from BR days then travelling with them as we were pushed through the tunnel on a scaffold mounted on a rail trolley. As we travelled through the tunnel at low speeds, 2 examiners and I would check the security of the brickwork by tapping with sounding hammers or poles and additionally I would be looking out for fractures and bulges in the brickwork. At the end of the examination a detailed report was produced which I would sign off.

Inevitably in work of this nature there was a lot of banter, some interesting results and a few amusing incidents. During BR days, I was almost badly injured by my High-Visibility vest! In electrified areas, the overhead equipment is always a hazard, and we were examining one of the tunnels near King's Cross and being pushed merrily along by the men below when my Hi-Vi vest caught on a dropper from which the contact wire was hung. Fortunately, I was at the front end of the platform because I had to walk along the length of it whilst frantically shouting stop, which they did just in time before I was swept off at the back.

There were also occasions when we came across brickwork that was so loose that there was a danger of it falling out. We always carried wooden wedges and the normal procedure was to wedge the brickwork to prevent it falling and make an urgent recommendation for repairs to be carried. In Railtrack days it was disconcerting to find that my recommendation for urgent repairs had still not been carried out by the time I returned a year later to carry out the next exam! On just one occasion, the state of the brickwork in a tunnel in Yorkshire was too bad to re-open the line without temporary repairs being carried out. It was a shallow tunnel which meant there was not much cover between the tunnel brickwork and the road above. Unfortunately, the road drainage was defective, and water had been running into the tunnel, eventually causing the brickwork to be so loose that about 2m2 fell when I sounded it with the hammer. This was far too much for the resources that we had on site. In BR days I would have had authority to call out my own staff and we would have sorted the problem very quickly and certainly before the possession was due to end. However, Railtrack control had to find a Railtrack engineer who then had to get authority to call out the maintenance contractor who had to find some men and materials, etc., etc. All this time I had to stay on site because the possession could not be given up until I was satisfied that the tunnel was safe. It took hours!

Since I travelled independently to the tunnels to be examined, I could get caught out with transport arrangements. Bramhope Tunnel is between Leeds and Harrogate and is VERY long – 2 miles 241 yards to be precise. Access is normally from Horsforth Station but, unbeknown to me, there is another access point at the opposite end of the tunnel. I met up with the examination team at Horsforth and after a bit of a delay, we started to examine the down side of the tunnel. I should note here that tunnels over about 2 miles long are usually examined over 2 visits, one for each side and that was the plan in this case. When the exam was finished, my colleagues told me that they were carrying on north to get all the equipment off the line at the next access point but since my car was at Horsforth, I set off back through the tunnel, having obtained their assurance that the possession would not be given up until I advised them that I was out of the tunnel. They set off north and I set off south but within 5 minutes, my handlamp failed and I was faced with a dilemma: do I carry on through the tunnel or turn back to try to follow them to the other access? Trouble was they might have already left that when I caught them up – then how would I get to Horsforth? In the end, I pressed on using only the light from my mobile phone screen reflecting off the steel rails to guide me (no torch facility on phones in those days). It was a very long walk-in absolute darkness!

The remoteness of some tunnels and the quiet darkness of the night sometimes results in seeing amazing natural events and sometimes some real frights. While waiting to start examination of Hindlow Tunnel, up in the

Derbyshire Hills near Buxton, I was treated to the most amazing meteor shower which was made special by the lack of light pollution. In Corby Tunnel in Northamptonshire, I have seen huge sheets of ice hanging from the roof of the tunnel. These had to be knocked down by men with poles before the line could re-open. On the way along the farm track that leads to Stoke Tunnel one night I was surrounded by a whole family of the tiny Muntjac deer that roam wild in the area. There are also strange happenings that set the imagination running wild. When I walked out of Thurstonland Tunnel near Huddersfield at about 4 o'clock one morning, my blood ran cold as first of all the atmosphere changed from cold to suddenly being to exceptionally warm and oppressive and then I heard a piecing scream. I thought maybe werewolves but then rationalised and decided that the change in temperature was because I had come out of the tunnel and into a sheltered cutting and the scream was probably a fox. Had me scared though!

There are 14 tunnels on the Settle to Carlisle line and I have examined every one, many of them more than once. Weather conditions in that part of the world can be pretty severe on a winter's night but the scenery is amazing. It can be very cold – I once registered -9 degrees on the car thermometer – and there has often been snow lying on the ground, though, I only once experienced a snowstorm and even that didn't last long. Some of the tunnels are quite short and we would do several of these together from a powered inspection platform which travelled from tunnel to tunnel with us in the "basket". Towards the end of my career, I was doing more training and certification work, and I was approved to certify Bridge Examiners and Tunnel Engineers. This led to some interesting site visits, particularly for companies who carried out examinations of closed lines. At privatisation, those assets that were not required by Railtrack for the operational railway were vested in a publicly owned organisation known as BRB (Residual). Their task was to sell off as much as possible and keep the rest safe, including a large portfolio of bridges and tunnels. Various contractors were employed by them to examine their structures and I would often be requested to certify their staff. I recall doing this for one team on a long tunnel in the Mendip Hills which had been part of the Somerset and Dorset railway. On one occasion, they also set up a special examination of a tunnel on the privatised "Bluebell" line in West Sussex. That was great fun because we had to wait for a steam train to pass through before we could take possession. Overall, I can say that I enjoyed my tunnel work and felt that over the years I had earned the right to be considered a bit of an expert on brick-lined tunnels.

Chapter 12

Bridge 79, Loughborough

Like Lady Elton's bridge, this reconstruction deserves a chapter of its own and it is an appropriate place to recap on the structure of the British Rail Civil Engineering Department at the time. At the very top was the Chief Civil Engineer at the BRB who headed a relatively small staff that were mainly concerned in producing standards that would apply to all the regions. The main thing that I knew them for was the production of Standard Bridge Drawings that could be adapted to fit most situations. Even these were often originally produced on the regions, for instance, steel box girder designs were from the Western Region and standard steel footbridges were a Midland Design, however, the BRB was the authority that approved them for general use. Reporting to the BRB, each region had a Chief Civil Engineer with a full technical office capability. These offices looked after major new works projects and had their own site supervisors. They also provided designs for the reconstruction of bridges which were often an adaption of standard drawings to suit the size and layout of a particular bridge. Down at the bottom of the pile were the Divisional Civil Engineers who were mainly concerned with repairs and renewals of both Track and Structures with a heavy emphasis on site supervision.

On the works side, in particular, we carried out all the designs for repair work but relied on the regional office to provide us with drawings for new bridges. However, in the case of the reconstruction of bridge superstructures, only the drawings for that part of the work were supplied to us and any additional repairs were our responsibility to design. So, if as was the case with bridge 79, one of the piers was in poor condition, then it was my job to come up with a repair scheme. Our approach was pragmatic and generally involved restoring the status quo or using "rule of thumb" principles to provide an improvement backed up with some basic calculations. All the design work at whatever level was carried out "in house" by experienced teams who were allowed to follow their own judgement. For instance, it was not considered necessary to prove the stability of existing bridge abutments if they had already been supporting the old bridge without problems for maybe a hundred years. Once privatisation took place and Consultant Civil Engineers and contractors were employed, everything had to be proved to meet current standards and this could result in very substantial extra cost.

Later in my career, my company, Bridgeway, was once paid twice for the same job despite protesting to the client that the work was unnecessary. A bridge in Lincolnshire was proposed by Railtrack for reconstruction of the superstructure using the existing abutments. It was clear that these abutments had stood for over 100 years and were still in good condition, but the consultants instructed us to prove the size and depth of the abutment foundations by drilling, which we were pleased to do. The job was then postponed and then a couple of years later reinstated but with the design work going to a different consultant. However, the results of previous drilling were not available to the new consultant for some contractual reasons, so we did all the work again. Thanks for the money, but no way to run a railway!

Bridge 79 is located just north of Loughborough on the Midland Main Line. If you are travelling across it by train, it is unlikely that you will notice it, even though it has 9 spans and would therefore qualify as a viaduct. It is only a couple of metres above ground level and each span is about 6m long. The job was to reconstruct the entire superstructure and carry out whatever other repairs were needed. It was the most time-consuming job that I can remember, and it had more than its fair share of incidents. The bridge carries the four tracks of the Midland Main Line over a small watercourse called Summerpool Brook and meadows that are prone to flooding. The original bridge was formed of brick abutments and piers with wrought iron main girders and a timber deck which supported normal ballasted track. The wrought iron beams had deteriorated, and parts of the timber decking were rotting so a decision was made to reconstruct the superstructure of the bridge thus removing the problems of maintaining a timber deck under track ballast. The brick piers were to be re-used but capped with new concrete cill beams.

When I first looked at the bridge, several problems became evident. The drawing supplied to us for the new structure showed no details of the piers and an inspection showed that one of them had been sinking and was badly fractured so I would have to design remedial works for that which would need to be completed before the main works could take place. As I have said, one of the advantages that we had as BR engineers was that we only had to answer to our own superiors for our actions and if they were satisfied, we had no need to prove our engineering judgement with endless investigations and calculations. As a rule, when dealing with local foundation problems such as the one at bridge 79, I adopted a simple approach which I thought was self-evident.

Firstly, this bridge had been in place for well over 100 years and of the 8 piers, only 1 was showing problems and even that had not reached a critical stage. Therefore, I reasoned that the existing foundations were not greatly overloaded so all I really needed to do was fully repair the pier and enlarge the foundation area, thus reducing the load on the supporting ground. I decided to double the width of the foundation, thus halving the pressure on the ground and then fully tie the pier and the foundation extensions together with steel reinforcing bars and grouting. This was a simple operation that our own staff could carry out. Such an approach would not be possible today because of the need for consultant civil engineers and contractors to fully prove their design calculations for fear of repercussions in the event of any problems.

We arranged to get access to the site via a canal-side private road belonging to the British Waterways Board and then across a field to the bridge. All went well with the pier repairs with only the occasional spraying of everyone around with cement grout under pressure until one lunchtime when the foreman decided to send the apprentice for fish and chips. The only transport on site was a small dumper with rear wheel steering which can easily catch out the unwary driver. The dumper could not go on the public road so the apprentice, a lad of 17, set off at great pace across the field, along the canal-side track and onto the towpath until he reached Loughborough Lock where he lost control and drove straight into the canal, fortunately jumping clear before finishing in the canal himself. This created a whole host of problems. Firstly, the apprentice was not authorised to drive the dumper, and secondly, in order not to use the public road, he had gone beyond the vehicle track and onto the towpath proper where vehicles were not allowed. To add to that, the dumper was needed on site and not where it was, causing an obstruction to boats trying to enter the lock.

I was in the office when I got a call about the whole sorry tale, and it was clear that we would have to act quickly before matters got out of hand. I ordered up a small road mobile crane from a local firm and told the Leicester Works Supervisor to get to the site to meet it but then wait till I got there. In the meantime, I set off for site, accompanied by Gordon Issott to act as my attendant, pausing only at our diving stores to pick up a dry suit. The crane had arrived on site just before me and the first thing that I noticed was that there was a culvert under the towpath that was a part of the filling and emptying system for the lock. I issued STRICT instructions that the crane must not work over the culvert, and it was placed in a safe position. I got into the water and hooked the crane onto the dumper which was swung clear and placed back on the towpath. A fitter from the plant department had already been summoned and he quickly got the dumper started and drove it back to site. There had been no sign of British Waterways Staff so as far as I was concerned, it was job done and I shared a joke with the apprentice because we all thought we had got away with it.

I left site shortly afterwards and went back to the office, only to be met with another urgent message to contact the site again. It transpired that just as it was leaving, the crane got a punctured tyre, which had to be changed. The procedure for this was to put down the crane outriggers to lift the wheel off the ground while a new tyre was fitted. Unfortunately, the outrigger was positioned directly over the culvert which collapsed under the load. We had now used the towpath for road vehicles without obtaining permission from the BWB and caused part of the lock system to collapse, if they found out, the proverbial shit would hit the proverbial fan. I went back to site and got a bricklayer from the bridge, who, fortunately, could repair brick arches and bribed him with overtime if he could get repairs completed before we all went home. He worked mainly from the top while I assisted by crawling into the culvert with my dry suit on and in a couple of hours, we had made an acceptable repair and tidied up. The crane was off site without any further problems. I never had any contact with BWB about the incident so I assume they never found out and we even succeeded in keeping it secret from "spanner man" who would undoubtedly have had a field day!

While the pier repairs were going on, we made plans for the reconstruction itself and booked the necessary possessions. These were not too difficult to get because we planned that we would only work on 2 out of the 4 tracks at any time and make the other 2 available whenever they were needed for traffic. Despite that, we were only allowed to work from last thing on Saturday evening until 6am on Monday. I started to draw out a site map and noted that there was a high voltage CEGB line supported on pylons crossing the north end of the bridge. Trying to solve the overhead line problem, however, proved to be more difficult and it wasn't helped by my poor knowledge of how high voltage power worked. You might think that the big, nationalised industries at the time would have worked together in an atmosphere of co-operation, but this was not always the case. When I met with the CEGB to discuss what I thought would be a simple matter of switching off the power for the duration of the possessions each weekend, I was met with a blank refusal and when I said well, we could manage with just the bottom 2 sets of wire switched off because our crane does not have a long jib, I was almost laughed out of the meeting. They simply refused to consider meeting my deadlines and said that I might have to wait for years to do the work until they shut down the line for their own maintenance.

Then I got a lesson on the practicalities of what I was suggesting. Firstly, I was told about 3-phase electricity which in this case meant that whilst they could switch off the power to all 3 lines on one side of the pylons, it was impossible just to switch off the bottom 2 and leave the others live. Then they explained that to switch off the power some giant levers would have to be pulled at Ratcliffe-on-Soar power station and once that was done, linemen would have to come to site and bond out individual wires on each side of the bridge. As a parting shot, I was also given a warning that whenever it was done, British Rail would be expected to bear all the CEGB costs which would run into many thousands of pounds.

I left the meeting and as I drove back to the office, I began to ponder what the legal situation was because clearly, the railway had been in place before the power lines were erected so some sort of legal agreement must exist for their presence over the railway. Had the power lines been following a public road then the provisions of the public utilities and street works act (PUSWA) would probably apply. This is the legal framework for all utilities in the public road and if the railway wanted cables in the road surface to be moved, they would have to bear the cost. However, this was not the case here so it was probable that the CEGB crossing of the railway would be by "Wayleave" in which case, it was possible that it might provide the solution, so I contacted the BRB property board who really came up trumps. The wayleave stated that when required to do so by the railway company, the CEGB had to remove their cables to facilitate any works to the railway AT THEIR COST. I could have jumped for joy and couldn't wait to get on the phone to the CEGB and make my demands for isolation. To be fair to them, once they realised the legal situation, they couldn't have been more helpful. An engineer was assigned to work with me, and I got the feeling that he was relishing the chance to do something different to his usual job. He even arranged for me to visit a CEGB national control room to see exactly how they dealt with the distribution of power throughout the country. When the time came for the power to be switched off, if was all done efficiently although I was told, in confidence, that there was an almighty flash when the lever was pulled in the power station because it had not been moved for years.

Planning the actual reconstruction was difficult because there was not sufficient possession time to complete all 9 spans of the bridge in a single possession, even if we only worked on one track at a time, Furthermore, that in itself would be very difficult because of the layout of the old bridge with timber decking under the "six foot" areas. I had to find something radical otherwise the work could not go ahead, and I eventually hit on a plan. The layout of the old bridge did allow us to make a break under the "ten foot" which is the wide section between the Main and Goods lines so we would carry out all the work under the main lines first. The plan was that during the first possession, we would remove all track, ballast, and timber decking, leaving just the main girders in place. This work could be done with small rail cranes and other rail-mounted plant working from the goods lines during Saturday night while there was no traffic. Panels of track with timber sleepers would then be laid on both sets of main girders and fixed directly to them with steel "U" shaped brackets. This work could be completed on the Sunday without affecting the goods lines by cranes working end-on. A 10 MPH speed restriction was applied to last while this temporary situation remained. This allowed us to work on the abutment tops between the girders and do other preparatory work and be ready for the next possession.

On the subsequent weekends, the old girders were removed, and the new cill beams and main beams were put in place and ballasted track installed. The schedule for this was tight but we felt that it was manageable. There were concerns about the security of the temporary brackets attaching track to beams so they were regularly checked and had their bolts tightened. Craneage for the work was the Toton 75-ton breakdown crane but it would have to work on the bridge itself when placing the beams. For this to be safe, the front outriggers had to be positioned over the tops of a pier so that the main loads were not carried on the old girders. The piers were built up to the level of the outriggers with giant timber grillages that could be lifted from one pier to another. The main crane was also supported during Saturday night by a 15-ton crane working along the goods lines which was used to remove main girders and to place beams along the "ten foot". Because the bridge did not cross any publicly accessible areas, the bridge was designed without waterproofing so that time consuming job was avoided, and it was this that allowed us just sufficient time to replace all the spans within the possession time.

The plans were approved, and the first weekend went ahead as planned. The temporary track arrangements were installed within the possession time and the main lines were reopened at 10 MPH. I was on site most days during this period and took a close interest in the performance of the brackets. All seemed to be well and one day I got absolute proof that they were up to the job. I was on the bridge, standing clear as a passenger train approached and realised that it was going much faster than the speed restriction allowed. It was an express that had stopped at Loughborough Station but under full power, it had reached at least 60 MPH as it approached the bridge! We all held our breath, as, to make matters worse, the driver realised his error and made a full brake application just as he hit the bridge. So, the train not only crossed the bridge at high speed, but the structure was also subjected to extreme braking forces. The train didn't stop, and we made a quick inspection of the brackets but thankfully found them all secure and in place, much to my relief. I didn't want to get the driver in trouble, but I was still fearful of what might have happened, so I spoke to the local movements inspector who arranged for special reminders to drivers to be posted on depot notice boards. I also got the local P-Way Inspector to carry out daily checks on the speed restriction marker boards. I think that mistakes like this were not altogether uncommon because sometime later a system of "reminder" boards for drivers was introduced if a station was located between the initial warning boards and the site of a speed restriction.

During these weeks, my daily routine was to go into the office at Nottingham first, then catch a train to Loughborough and walk down the track to site where I would stay till late afternoon and then go straight home, again by train. Since we only had one van between all of us in the bridge section, this avoided tying it up on the same job every day. It was summer, however, and I soon hit on a better way of doing the journey. I would take my bike on the train to Loughborough, then cycle down the cess to the site (I had a special pass for that) and at the end of the day, I would cycle the 20 or so miles home. This was really the start of my love of cycling which persists to this day.

The cill beams for the tops of the abutments and the parapet beams for each side of the bridge were pre-cast concrete and were being manufactured for us by a company in South Wales whereas all the other main beams were of pre-stressed concrete which were manufactured by a different company who were located at Rye House in Hertfordshire. Both manufacturers still had sidings in their premises so the plan was to load them into wagons and send them by normal freight train to Toton yard where they would be marshalled up into the ballast train for the job. At the time, BR's wagonload business was undergoing a transformation which it was hoped would revive it and the old vacuum-braked trains were being gradually superseded by a new service called "Speedlink" which used air-braked wagons that travelled much faster and kept to specific schedules. "Speedlink" had been already introduced on the services that would bring the pre-stressed beams from Hertfordshire but not yet on the services from South Wales. This meant that our ballast train would be made up of both vacuum and air wagons and therefore would have to come to site under the control of the engine brake only. Not really a problem because of the short distance involved, at the time there were still ballast trains which ran un-braked.

I looked up the wagons that we needed and ordered up OBA wagons for Rye House and "Pipe" wagons for South Wales. The freight staff in Nottingham made the necessary arrangements with their counterparts that looked after the 2 sidings, and I even took a trip to Rye House to discuss the loading of the beams with the local inspector. He was very interested because it was the first time that the "Speedlink" service would be used at that yard. As the weekend when we needed the first deliveries approached, I kept a track on the wagonloads of beams through the excellent TOPS system which allowed me to see exactly where they were at any given time. The wagons from Rye House were moved down to a yard in London and from there an overnight "Speedlink" service brought them to Toton in good time. The wagons from South Wales, however, stalled when they got to Severn Tunnel Junction and when they were still there on the Thursday before the weekend I began to be concerned. In Furlong House, the Divisional Control was situated directly below our office so I popped in there to see what could be done.

Having located the freight controller, I explained the situation to him, and he said that he would do what he could. However, by Friday morning nothing had moved so I went back to control. As usual, they were very busy moving coal trains about to suit the requirements of the power stations, but I managed to get a hearing and was told to come back later. I returned during Friday afternoon, and I really was beginning to worry, it was only about 36 hours before the possession, and we still had no beams for the bridge. That was when the controller put an arm round my shoulder and told me not to worry. "Brian," he said, "leave it to me. I am on duty tomorrow and there will be no bosses about to object. There are enough wagons in Severn Tunnel Junction to justify me organising a Control Special train so you will get your beams."

With that, I had to put my trust in what I had been told but if that failed, I would really have egg on my face and there were those in the office who would say that I placed too much faith in the railway system. Anyway, there was nothing more that I could do so on the Saturday afternoon I went to the football match in Leicester. I made this journey by train and the walk back from the ground to the station took me alongside the line. It was then I heard a freight train passing and looking over the wall, I was greatly relieved to see my beams trundling past on their way north. The only thing that I had to worry about now was getting them marshalled into the ballast train when they arrived in Toton but my friend in control had all that covered. The train was stopped at Loughborough and the wagons were detached there and shunted into the yard so in fact they arrived before the rest of the train.

With that worry over, the actual reconstruction continued more-or-less as planned but it did look strange with the crane operating on what was a skeleton bridge with its outriggers supported on 1.5m high timber grillages. As usual, I did first and last shifts with one of my bridge section colleagues covering the middle shift. When I left site, we were a little behind schedule but going OK and when I returned, we were still progressing well but a little further behind. Possession of the two main lines had to be given up at 6am which was about the same time that the first London train of the day left Derby and as that time approached there was a decision to make. We had caught up some time and by 5am, the P-Way were busy completing the laying of the tracks with at least an hour's work left to do. The operations department had already been in touch, and I had to decide if the 6am from Derby could pass on the main lines or if it would have to use the goods lines. This may not seem to be a major problem but if it had to be diverted to the goods lines, the train would be delayed, and it would count as a possession over-run. I would have to explain personally to the Divisional Engineer, who in turn would be called over the coals by the Divisional Manager. On the other hand, if I allowed the train to approach on the main line, and it had to wait while we finished work, then the delay could be more serious, and I would be in trouble for that. In the end I took the risk and told them to send it down the main lines. The supervisor prepared himself for a quick hand back by getting his men ready to remove the detonators from the line as quickly as possible and although we didn't actually finish till about 6.10am, we did not delay the train at all, and no over-run was recorded.

The remaining weekends of work went fairly well to plan and in the end, I felt that we had done a pretty good job, but there was one more incident to contend with before the job was completed. There was a junction between Loughborough Station and the bridge where trains on the main line could be crossed over to the goods lines and vice versa. The signals controlling the junction were set high above the track so that they could be seen by the driver even though there was a road bridge in front of them and because of this, they could also be seen from the bridge site.

Standing on the bridge during one of the possessions, I happened to look up and saw to my consternation that the signals were "pulled off" for the main line which would send and train directly into the possession and possibly into the hole where the bridge should have been. I shouted to the operations supervisor who was with us, but he had already seen the problem and he was running towards the nearest signal post telephone. By this time, I could hear the train leaving Loughborough Station and began to fear the worst, but I left it to the operations man to sort it out. He was running towards Loughborough to be ready to stop the train if necessary, but the train came to a stop at the red flag which was placed on the track at the time of taking the possession (a STOP board is now used instead of a flag). The possession was also protected by detonators placed on the line so there was no real risk of the train reaching the bridge. The operations manager then had the task of arranging for the train to be set back into the station so that it could resume its journey on the proper line. I don't know if there was an official enquiry into the incident, but I suspect not, although I am sure that the signalman got a roasting and probably an official reprimand. It did go to show the value of placing physical reminders (in this case, red flag and detonators) on the track. This is a process that continues today, although I am sure, now that the same protection could be achieved electronically.

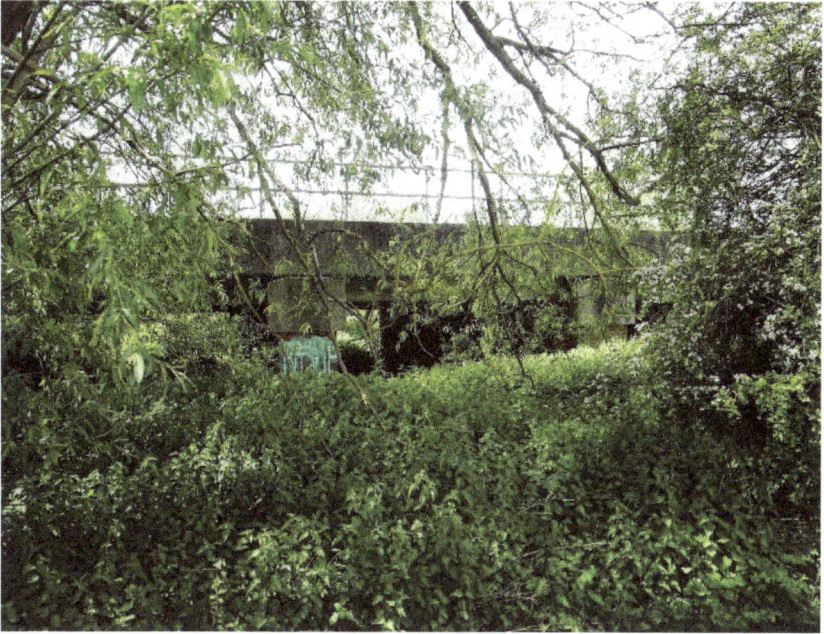

Bridge 79 – East elevation glimpsed through the undergrowth Photo - Brian Maddison

Loughborough lock – The dumper went into the canal just here. Photo Brian Maddison

The culvert that had to be repaired

Chapter 13

More Bridge Reconstructions

There is no doubt that carrying out bridge reconstructions was one of my favourite occupations. I have mentioned some of them in previous chapters but there are many more and I have grouped some of them together here. The first that I am including in this chapter was disguised as a bridge repair and I will explain why. I was in the drawing office at Nottingham one day when there was a bit of a commotion and we soon learned of an unfortunate accident on one of our bridges. An elderly gentleman had been climbing up the staircase of "Bobbers Mill" footbridge in Nottingham when one of the timber treads had collapsed and he had been badly injured in the fall. It transpired that the tread was rotten, so it was clear that the responsibility was squarely on the shoulders of our department. Rot in the ends of timber treads on metal footbridges was a well-known problem and could quickly become dangerous if not spotted and dealt with. It should have been picked up by the bridge examiners and acted on quickly by the local works supervisor but somehow, it had got missed.

The bridge was closed, and I was given the job of organising repairs. Once I had looked at the bridge, I realised that if I stayed quiet, it might be possible to do something rather special. I have pointed out before that at Divisional level we were not allowed to design our own bridges, but I also knew that the Chief Civil Engineers drawing office in Birmingham would not be particularly concerned about a small footbridge so if I could disguise the work as "repairs" rather the "reconstruction" I would be able to keep the job in our own office. Keeping quiet was not only so the Birmingham office didn't know too much but also to get as much as possible done before "spanner man" got wind of the project and did as his name suggests.

The bridge was supported on brick piers which were in fair condition and the 2 main girders were sound, but the staircases were very poor, and the timber deck of the bridge had some rot. In view of the accident, I got an agreement that replacing the deck in timber would be a bad move, so I was instructed not to use timber in the repairs at all. That gave me the go-ahead to design new staircases and new decking whilst preserving the illusion of a repair job by re-using the old girders and the brick piers. I got busy on the design work which involved not only much work with the slide rule but also several visits to the Nottingham University Library to look up design standards which were not held in our own office. I then had to produce the pen and ink drawings which included a great deal of steelwork detailing since they had to be used by the steelwork manufacturer. The one problem that I had was how to arrange for the manufacture of the new steelwork which would require the expertise of a good manufacturing organisation. There was no capacity at Nottingham to prepare contract documents that could be issued for competitive tender and passing that task to Birmingham would have lost us control of the work so I looked around for what "in-house" capacity might be available within the railway family.

As luck would have it, Derby loco works were known to be short of work to the extent that closure was being forecast so I approached them to see if they could do the work. As you might imagine, they were most enthusiastic and did an excellent job for us. The old girders were sent to Derby loco where they were refurbished and fitted with a steel deck. They also manufactured the staircases and other bits and pieces. I went over to view the work in progress and found that aside from our bridge, there was not much work being done and it was no surprise that before long, the whole works closed. Eventually, the new bridge was ready and moved to site by train. Lifts were fairly light so instead of a breakdown crane, I elected to use a 15-ton capacity diesel crane that normally worked in Lenton Yard handling panels of track. That proved to be a bit of a mistake that nearly cost me dearly. The driver of the crane was known to be somewhat cavalier and when the time came, he sped along the track towards the bridge with the main span hanging from the hook. Unfortunately, I was standing on the top of the pier waiting to guide the bridge into place and I had to grab hold of the bridge and hang on to avoid being swept off the pier and falling 20 feet down to the track below. After an exchange of expletives, we eventually got the bridge in place but not without difficulty and to be fair to the driver, the movement of the old crane was difficult to control with any accuracy.

Once finished, the bridge looked great and it still looks good today although it is covered in graffiti. Just after the bridge had been installed, I got a visit from Les Wood from the Chief Civil Engineer's office, and it became obvious that Birmingham had been tipped off about what we had done, and I strongly suspected "spanner man". I had to take Les to site, and he went over all my calculations and drawing as well as doing a detailed inspection of the bridge. Fortunately, he could only find one thing to complain about. I had used site welding to secure small "keep-strips" at the bearings after the bridge had been placed. Unknown to me, this was against BR policy but since they were not structural welds, they did not require testing so Les decided not to insist on any change. In fact, Les was a thoroughly decent bloke and I think he was embarrassed by the policing role that he had been given.

Bobbers Mill Footbridge in 2020 – suffering from graffiti but otherwise unchanged

Another similar job began, strangely, with the conversion of the level crossing at Meadow Lane, Long Eaton, from wooden gates operated locally from an adjacent signal box to lifting barriers controlled from Trent power box. There was a wrought iron footbridge at the level crossing which allowed pedestrians to cross the line when the gates were closed. It spanned 4 tracks and was maybe 60 feet long. It was deemed that it could be closed and to give room for the new barriers, I got the job of removing it. It occurred to me that the footbridge was in good condition and that it might be re-used for another project that was in the programme of works for the year. I therefore arranged for the bridge to be taken down carefully, including the staircases and transported to Sandiacre yard where it could be worked on. The job that I had in mind was to use it to replace a footbridge on the Wirksworth Branch, just outside Duffield in Derbyshire. This line was used to transport stone from quarries around Wirksworth down to Derby for onward despatch around the country. The existing timber footbridge was in very poor condition but with a relatively small amount of work plus some minor repairs, the Meadow Land footbridge was made to fit. The location for the bridge was in a cutting and it had to span from an existing brick pier on one side to a new bank seat on the other. One of the staircases from Meadow Land was adapted to fit to the pier side of the bridge.

It is worth noting at this stage that our Derby depot had a number of what might be described "characters" amongst the site staff. One was the lorry driver named Bert (I don't recall his surname) and another was a former refugee from Eastern Europe called Joe Bernans. Bert was known to run several side lines, one of which collecting old timber which he would purchase on "Firewood orders" and then using it on his own smallholding or selling it on at a profit. The latter practise was forbidden but since it got rid of scrap timber that the railway company might otherwise have had to pay to dispose of, then no one saw fit to make a fuss. In fact, it was well known that when it came to dispose of anything unwanted, it was best to leave the job to Bert and ask no questions. Of course, this was long before the regulations were imposed regarding the safe disposal of waste, but we knew that Bert would find an appropriate home for the material and that it would not be fly tipped.

Joe was a different kettle-of-fish. He was an incredibly hard worker and once given a job, he wouldn't stop for a break until it was complete. The only problem was that the job would not necessarily be done in the way that had been originally intended. When I first knew Joe, he had a small house and a lady that we all assumed was his wife. However, his home life was somewhat turbulent and occasionally, he would bed down in the carpenter's shop for the night when she had thrown him out. Eventually they split altogether with Joe openly saying she was a "Dirty woman" that he would have no more association with. I was faced with deciding whether to let him stay living in the carpenter's workshop or evicting him, which would also probably mean him losing his job. In the end we made him a semi-official and unpaid security officer, thus allowing him to stay on the premises at night. Joe had a passion for lighting fires on site using any wood that he could get his hands on and liberal amounts of petrol, which were generally welcome on a cold winter's night, but which had been known to get out of hand.

We had a long weekend possession of the branch to remove the old footbridge and install the new. The old footbridge was taken down piecemeal by hand and as usual, Bert had negotiated to buy all the timber which he was to be allowed to remove from site of the next few days. However, once the timber began to be piled up on the top of the cutting, Joe stacked it up and set light to it. I told him to put it out, but he just walked away grumbling so a couple of us extinguished the flames before they really caught. I told Joe not to light it and then added that it had all been bought by Bert. It transpired that was a mistake because Joe and Bert didn't get on and the last thing Joe wanted was for Bert to make a profit out of selling the timber. We brought in the steam breakdown crane from Toton to install the new bridge and everything fitted very well. Once I was sure that all was well, I left site, leaving others to finish tidying-up. Apparently, as soon as my back was turned, Joe poured petrol onto the timber heap and set fire to the lot! When Bert arrived later for his first load, all he found was the smoking remains of a bonfire! I am not sure whether to two of them actually came to blows in the ensuing argument, but I doubt it because Joe was as strong as an ox and would have made mincemeat of Bert. I decided to stay well out of it because I knew if I took any formal action, it would have uncovered all sorts of secrets that were best kept hidden.

The line was eventually shut by British Rail but in the 1990s, a group of enthusiasts known as Wyvern Rail was formed with a view to operating a heritage railway over the line and in 2011, they began operating a full service between Duffield and Wirksworth. It is now called the Ecclesbourne Valley Railway and the station in Duffield is adjacent to the Network Rail platforms. The footbridge remains and has been well-maintained and re-painted by the new railway company. I took a trip on one of their trains just for the fun of it and also cycled to the footbridge to pay it a visit.

Top of Duffield footbridge

From the train

Staircase

The train in Duffield station

Ashby de la Zouch is a sizable town on the Leicester to Burton Line. This line remains open today for the carriage of freight, mainly aggregates from the quarries at Bardon Hill and Stud Farm. It last carried passengers in 1964 and more recently it was very busy with coal from several collieries along the line, much of which went to the massive power station at Drakelow. To my mind, the line has to be a prime candidate for passenger re-opening, serving, as it does, several large towns with a combined population of over 100,000. During the 1980s, the bridge carrying the railway over Station Road in Ashby town centre was in need of reconstruction, a job that I was very pleased to be allocated. The bridge was a single span wrought iron structure which had been severely damaged over the years by being struck by high vehicles. The proposed new bridge was a standard Western Region design of steel box girders. This design had been adopted by all regions of British Rail and it consisted of 3 trapezoidal box girders for a 2-track bridge that supported steel deck panels with integral welded cross-girders. These had to be fastened on site to the box girders with cold swaged rivets known as "Huck Fasteners". Normally, there were steel walkways fastened to the outer box girders but for this bridge; they were replaced with independent concrete walkway beams that also acted as crash protection to stop damage to the main structure from high road vehicles. Of course, I selected the Toton breakdown crane for the task which by this time had been converted to diesel. The reconstruction procedure was to first remove the track then lift out the old bridge and load it on to rail wagons to be taken away. Concrete cill beams were then to be installed at each abutment to replace the old bed stones and the new bearings placed in position.

Great care was needed in setting out the bearings as there is little tolerance in the alignment of the bolt holes. I had already learned this when I had visited the factory that fabricated all the steelwork. It was interesting to watch the bridge being put together in a trial erection that was done to prove to us that everything fitted together but I noted then that the factory was able to employ a range of jacking system to line up the bolt holes which would not be available to us on site! Each of the 3 main girders were to be lifted off their rail wagons and positioned on the bearings; each section of the deck was then offered up to the girders and had to be held in position by the crane while all holes were aligned and temporary bolts installed. If the main girders were not correctly positioned, the deck units had to be placed back on the wagons and the crane re-attached to the main girders to move them. This was a time-consuming task to be avoided at all costs and if required would be a black mark against my name in the eyes of everyone else on site. Finally, when all the steelwork was in place, the whole deck had to waterproofed and the ballast and track installed. The walkway beams could be installed at a suitable time in the possession.

The possession was taken on a Saturday morning, after the passage of the last overnight coal train and had to be handed back late on Sunday night because coal trains had to be on the move again in the early hours of Monday. I took the first 12-hour shift and arrived early on site to make sure there were no delays to the start of the job. All went well, initially, the tracks were quickly removed, and we were just preparing to start dismantling the bridge when I received a message from control saying that the reconstruction might have to be cancelled. Apparently, the guard for the ballast train that was bringing the bridge units to site from Toton yard had checked his paperwork and found that there was no "out-of-gauge load" form that would authorise him to proceed with our load of girders. This was despite the move having been agreed with the operating department some weeks earlier. Realising that unless I could provide a compelling reason to continue, a very costly postponement would be the easy way out for control, I decided to force the issue and I told the steel gang to start cutting up the deck of the old bridge, knowing that once that was done, there would be no way back. I then rang control from a lineside telephone and told them in all truthfulness that the cutting up of the old bridge had already started and if the line was to reopen at all, they would have to get the new beams to us.

I could almost hear a pin drop at the other end of the line, but the controller calmly told me that he would get back to me when he had resolved the issue. Fortunately, for this job, the crane came to site on a separate breakdown train so when it arrived, we completed the dismantling of the bridge and the preparation of the abutments while we waited for news of our new beams. I knew that the guard would not take the train out on the say so of a telephone call and communications were not as easy as they are today so I wondered how control would solve the issue. Eventually, they got back to me and said that it was all sorted and the train was already on its way to site. Control had contacted the regional duty operations manager who was based a Crewe. He had to go into his office, verify the paperwork for the out-of-gauge load and then type the whole instruction into the "TOPS" machine so that it could be received at Toton and put into the hand of the guard. Since the TOPS machine was principally designed for the transmission and recording of wagon numbers, this was quite a prodigious feat for which I was most grateful.

By the time the train arrived, we were all ready to start building the new bridge. This operation went reasonably well, with the crane positioned close to the abutment, the lifts were well within capacity. Ensuring that the abutments are correctly prepared and cill beams installed in the right position involves a lot of climbing up and down from one side of the bridge to the other, so it is always a relief to get the first main beam into position and then run across it as a sort of traditional celebration. Once the first 2 main beams had been placed, the deck sections between them were fitted. This is always a time-consuming job to get all bolt holes lined up and then the temporary bolts installed. To do this, one man has to be inside the box girders to push the bolts into place while his mate on the outside ensures that holes are aligned, fits washers and nuts then tightens up. All went well, if a little slowly, with a great deal of hammering and swearing. Eventually, all steelwork was in place and the temporary bolts could be replaced with HUCK fasteners. Each bolt is removed in turn and replaced with a HUCK pin which is pushed in from the outside. The man inside the box girder, lying on his stomach in a very tight space, then fits a collar and applied the HUCK gun. This is driven by hydraulic power and the gun grips the pin and applies pressure to tighten the joint whilst also swaging on the collar. Once the correct pressure is achieved, a narrow neck on the pin breaks with a very loud bang and the joint is made. This is one time when there were no objections to being told to wear ear defenders!

Generally speaking, the rest of the job went well, and we were able to re-open the line on time. Throughout the daylight hours, we had an audience of townspeople and other onlookers, including, from time to time, members of the mines rescue team that was based near the bridge. Having completed my shift, I went home for a few hours rest before returning to site and staying there until the work was completed. Fortunately, although I was on site for a combined time of about 30 hours, there was no need to take food or drinks. The cook in the riding van attached to the breakdown crane could be relied upon to provide breakfast, lunch and dinner as well as copious quantities of tea. It was one of my most satisfying jobs, particularly when I reflect on my somewhat rash decision to cut up the bridge without knowing that I would get the materials train delivered to site.

The Toton 75-ton crane towers over bridge 55 at Ashby –de-la-Zouch

Bridge 55 in 2020

In 1978, I had the job of reconstructing bridge 31 at Blaby and I would have to say that it was probably not my finest. The bridge carries the South Leicestershire Line between Wigston Junction and Narborough (which is 2-track at this point) over the main road between Leicester and Blaby. Much as I wanted to use rail cranes on the job, it would have required 2 cranes doing a tandem lift that could be done by a single road crane at much less cost, so I really had no choice. The possession time was about 30 hours, from after the passage of the last service on Saturday night until first train Monday morning. This was not exceptional but was, nevertheless, a tight schedule and it was a point of principal never to hand back possession late. In order to minimise time taken to remove the bridge, I requested a crane that could lift off the old bridge in one piece then put it down on the road for cutting up before loading the pieces into rail wagons. Once the track and ballast had been removed, the ironwork was not particularly heavy, but the crane did have to stand close to the bridge. I discussed all this on site with a representative of the crane company some weeks before and all the lifts were agreed. The new bridge was formed of 2 "Z" deck spans, similar to those used at East Langton (Chapter 6) and these had been put together in Leicester for moving to site on rail wagons. Had I been using a rail crane, the actual crane supervisor would have come to a site meeting and would have been fully conversant with the job before the "day of the race" but only a driver is provided with a road crane and that proved to be a problem.

I had agreed to meet the first shift supervisor, Henry, at the County Arms Pub in Blaby about an hour before the start of the job. We had a couple of pints, but Henry was a drinking man and I think he had already had one or two before I arrived. Anyway, I got him out of the pub, and we took possession of the line on time. The crane arrived while the track and ballast were being removed but when I told the driver what I required him to do, he flatly refused to lift the bridge out in one piece. I showed him all the details of weight and radius and pointed out that the lift had been agreed with their company, but he was still not happy. He was not concerned that the bridge deck was too heavy but that it might fold up when lifted and that would be a problem for him because his cab would be very close and below the bridge when he lifted. I was cross at the time but on reflection, I could see his point. I just wished that either I or the crane company's representative had foreseen this before the start of the job because I could have planned accordingly and at the very least, I would not have needed such a large crane.

I accepted that the bridge would have to be removed a piece at a time and Cliff, the steelwork supervisor, got his men to work cutting up the deck.

Every cross-girder had to be cut at each end as well at the deck plates themselves and then the 3 main girders had to be lifted off separately. It cost us at least 4 hours delay but we had erred on the side of safety and the work was able to proceed. We finished removing all the old bridge by Sunday lunchtime and with no more major snags, the new bridge was in position, ballast placed, and track laid just in time to pass the first train on time on Monday morning. While placing the ballast, I had one more surprise. I should say here that standard "Z" deck bridges were designed for economy and were known to deflect under load because of their lightweight construction. Much different to the old bridges which were inevitably over-designed and even after years of corrosion deflected very little.

After the bottom ballast had been placed and the new track laid, we ran a train of ballast hoppers over the bridge so that the beds could be filled. The loco was a Class 47, quite large and heavy and I watched with interest as it crossed the bridge. There was a substantial deflection, to my tired eyes, it appeared to be an inch or more but in reality, it was probably much less than that. I checked all the bearings while the train ran the ballast out over the bridge and saw that they were all in good order and the beams were sitting correctly. I could find nothing wrong, and it had to be a combination of my imagination and the normal deflection of the beams. Even the next day I was worried, so I went back to the bridge and watched several trains cross the bridge and confirmed that there was no abnormal deflection. The bridge is still in place after over 40 years so I guess my fears could be said to have been unfounded.

South Leicester Line – Bridge 31 at Blaby. Preparing abutments while the crane waits. My father-in-law who was a photographer for the local paper took this picture.

Bridges come in all sizes of course but even small ones have their own challenges. One year while I was in the bridge section, I was allocated a small farmers underbridge (known as a cattle creep) to reconstruct. It was located on the Leicester to Peterborough line near Frisby. The old bridge was of wrought iron girders supporting longitudinal timbers to which the rails were fixed directly and therefore there was no ballast and no sleepers to remove, and the new bridge was to be a simple concrete beam structure. The only trouble was that because it was a small bridge, no provision had been made in annual possession plan, so I was limited to what was known as a "No Trains Period". Possessions for small jobs are arranged when there are no booked services, anything that requires a longer possession is fitted into an annual plan which will mainly accommodate track relaying work and major bridge work. Depending on the line, these were normally between about 16 hours and 30 hours but might require diversion of services and possibly bus substitution so comparatively few were allowed, and all work was planned well in advance. These were known as "Rules of the Route" possessions. Anything longer was almost unheard of which is a far cry from today when it seems the norm to close large sections of main line for the whole of Easter, Christmas, or every other Bank Holiday. I am amused by statements and press releases that say that such lengthy closures are necessary "for the safety of staff". In reality they are caused because the railway no longer has the in-house capacity to do the work and contractors, who may or may not be experienced in railway work, will demand long periods without out any trains and make sure, in order not to be hit with over-run penalties, that they have plenty of time to spare. In BR days, we didn't have that problem, put whatever work we did have, had to be fitted into the agreed "Rules of the Route" possession. What we did have to our advantage was the ability to improvise. For instance, if a job could not be completed before the first train was due, we could pass the train at a reduced speed and then complete the work afterwards. Such arrangements are not available today and I also acknowledge that modern signalling can often take much longer to reinstate and test after completion of the work and before trains can use the line.

To get back to Bridge 11, the last train between Leicester and Peterborough in those days passed at about 10 o'clock on a Saturday night and the first train on a Sunday was not until after 9am, so there was about 11 hours available to carry out the bridge reconstruction. This was the shortest period in which I had ever attempted such an operation, but on the other hand, it was also a very small bridge. There was no road access to the site, so the usual breakdown crane was to be used to lift the old bridge out and put the new one in. Once all that was done, a short ballast train would bring in the rails, sleepers, and track ballast required to complete the job. Possession was taken after the last train passed and the breakdown crane was then propelled to site from Syston Junction end. Having been told that the crane was on its way, I went a small level crossing near the bridge and waited for it to arrive. There I waited and waited for what seemed an age until it eventually arrived with an operations inspector walking in front. I have no idea why he thought that was necessary or what rule he had invoked that told him to do that, but I reckon that we were delayed by about an hour waiting for him to arrive. After that, there were no more problems and in truth, we finished the job with plenty of time to spare before the first train on Sunday morning.

The next bridge "reconstruction" that I was involved in was the Great Central Railway bridge over the top of Nottingham Midland Station. The Great Central Railway closed as a through route from London Marylebone to Leicester, Nottingham Victoria, Sheffield and Manchester in the 1960s and by the time I moved to the Leicester office, the line had been truncated and the only section remaining open for passengers in our area was between Rugby, Leicester (Central), Loughborough (Central) and Nottingham Arkwright Street. There was also a connection open at the Nottingham end for freight from the Midland line to Weekday Cross junction where trains reversed to run south over the Midland station bridge. The line North of Weekday Cross was closed to accommodate the building of the Victoria shopping centre. All this was as a result of the "Beeching" policy of line closures and whilst not all of the blame can be laid at the feet of that gentleman and it was clear that many branch lines were no longer needed, the closing of major routes simply because they were seen as "duplications" was undoubtedly a disastrous policy. At the very least, the track bed should have been protected for the future. The result is now that the railways are very short of capacity, and we are trying to build new lines such as High Speed 2 at enormous cost and immense disruption to the public.

By the time that I moved to Nottingham, the passenger service had ceased altogether but freight still ran, particularly to a military base at Ruddington and there were occasional trains of tanks and other military vehicles to be seen crossing the bridge. However, the reversal at Weekday Cross made this route difficult and it was decided to install a new chord line at Loughborough to give access to the line from the southern end. Once this was done, the line North of Ruddington closed and the viaducts at Nottingham across the River Trent and Nottingham Station were scheduled for demolition. The remainder of the line between Leicester and Nottingham was eventually severed at Loughborough by Alan Hughes who removed the bridge over the Midland Main Line (much to his regret, but he had no choice) leaving the Southern Section from Loughborough Central to Rothley to be operated as a very successful heritage railway and the northern section open for freight to East Leake and Ruddington. Once that ceased, this part of the route was taken over by a different preservation group who continue to run steam trains along the line and commercial freight trains of Gypsum from power stations to British Gypsum at East Leake via the Loughborough chord. There are now firm proposals to join the two sections together and a new bridge has already been installed over the Midland Main Line at Loughborough.

I was allocated the job of removing the massive viaduct over Nottingham Station which Comprised 2 long steel spans, one (the longest) crossing from Station Street to Platform 4 and the other from there to the car park. There was also a series of brick arches across the car park. I mentioned at the beginning that this job could be considered a reconstruction but there was some 35 years between my demolition of the old bridge and the erection of a new bridge at the same site by the Nottingham City Council for a new tram line. At the time I was given the job, the idea of trams using a bridge at this location would have been considered very far-fetched, but a new bridge has been placed there and it is both functional and very attractive. So, my job was "simply" the demolition of the old viaduct but imagine yourself being given this job and being told that it had to be demolished without closing the Midland Station at any time and that only on a Saturday night/Sunday morning would I be allowed to close any of the platforms and even then, at least 2 platforms had to be kept open for overnight trains.

This was the task that that I was presented with and the first decision that was to be made, was who would carry out the work. Following discussions with my boss of the time, we decided that our own staff and resources would be employed for all work that had a direct interaction with the operational railway but that the aim would be to provide a safe environment, separated from the trains and passengers within which a demolition contractor could work. The part over the car park was easy to decide upon, we simply closed off an area from the public and let the demolition gang work unabated but the two spans over the tracks would take more planning. Eventually, I came up with a scheme to provide a "crash deck" underneath the two spans which would be strong enough to support existing bridges and the demolition plant and once erected this would allow the demolition to take place piecemeal without affecting the railway below. Easier said than done, I think you would agree. The contract for demolition was awarded to a local company "Matrixgrade", who proved to be very good. Meanwhile I looked to the railway "family" to provide me with the resources for the crash deck.

My first port of call was to Newton Heath depot near Manchester where the Midland Region held its stock of temporary steel beams and bridging units and also stock of both "Light" and "Heavy" military trestling. Military trestling was developed, as the name suggests, by the Army and was used extensively during World War 2. Coupled with temporary "Bailey" bridge units, it was used throughout the UK to replace bomb-damaged bridges and throughout Europe and the rest of the world to replace damaged bridges that were needed for the allied advance. After the war the railway was allocated a considerable quantity of this material which was deemed surplus to military requirements, and it proved very useful. Fortunately, we held in the office a full catalogue of these units which included details of how to build them into full bridge piers. Working with these details I was able to produce a design which would carry the weight of the old bridges and, most importantly, could be installed under the bridges and still allow room for trains to path underneath. Once that was done, I went to Newton Heath with my shopping list but I found that even their extensive stocks could not provide all that I needed. For the remainder, I contacted my friends at Woodburn shops in Sheffield where much of the eastern region stock was held and in return for agreeing to make site work available for their staff, they provided the rest. At the time, between platforms 3 and 4, there were 4 tracks, one for each platform and two centre lines used for non-passenger traffic.

After much pleading, I was allowed to close one of the centre lines for the duration of the works and this was central to my plans. To support the deck under the largest of the spans, we erected 3 military trestling piers, one alongside Station Street, one on a centre line and one on platform 4. Once these were up, it was necessary to put up the beams which would span between the new piers. Since these had to go under the old bridge and be threaded around various other obstructions, including the station awnings, it was no easy job. As usual, I found salvation in a breakdown crane. By using the smaller of Toton's cranes, the 36-ton steam crane which had a shorter jib than most, we were able to offer the beams up to the trestles than winch them into position. The first attempt, however, almost finished in disaster for the steam crane. The beams were fairly light, so we did not need to use the outriggers for the crane and it was able to work "free on rail", fetching the beams from the wagons under its own power and lifting them into position. At the first attempt, in order to lift the beams high enough, the crane supervisor was watching the beams lifting and didn't notice that the hook block had gone up so high that it became wedged against the pulley wheel at the top of the jib, a condition known as "Block and Block". It became jammed and was only released after about an hour by members of the crane crew climbing up the jib and hitting things with very large hammers. After that, the job went reasonable smoothly and by the end of the second weekend, the crash deck was in place and full demolition could now go ahead.

Meanwhile the contractor was struggling with the amazing strength and hardness of the brick arches. The Great Central Line had been built much later than most railways and benefitted in having very good materials and a very skilled workforce. Their brick-built bridges are incredibly strong, and many remain in good condition today. Using every bit of heavy plant at his disposal, the contractor was making slow progress with this part of the work.

The steel spans were easier because although they were well-built, the early steel used at the time contained some impurities and was prone to rusting so some of the steel being cut up was already thinning. Eventually it all came down and we could remove the crash deck to leave a station now bathed in sunlight instead of the darkness created by the old bridge. I still have a memento of the job in the form of a small rockery in my front garden. These are broken pieces of bed stone from the bridge that I begged from the contractor. There the story ended, as far as I was concerned but in 2015, the Nottingham City Tram system was extended across the railway to the south and west of the city. A new bridge was built on the line of the old Great Central Railway, and I think this can be said to be the completion, albeit belated, of the reconstruction.

Great Central Bridge over Nottingham Station, prior to demolition.
Copyright Nigel Tout. greatcentral@nigeltout.com

Great Central Bridge over Nottingham Station, prior to demolition.
Copyright Nigel Tout. greatcentral@nigeltout.com

Great Central Bridge over Nottingham Station, prior to demolition.
Copyright Nigel Tout. greatcentral@nigeltout.com

Great Central Bridge over Nottingham Station, during demolition.
Copyright Nigel Tout. greatcentral@nigeltout.com

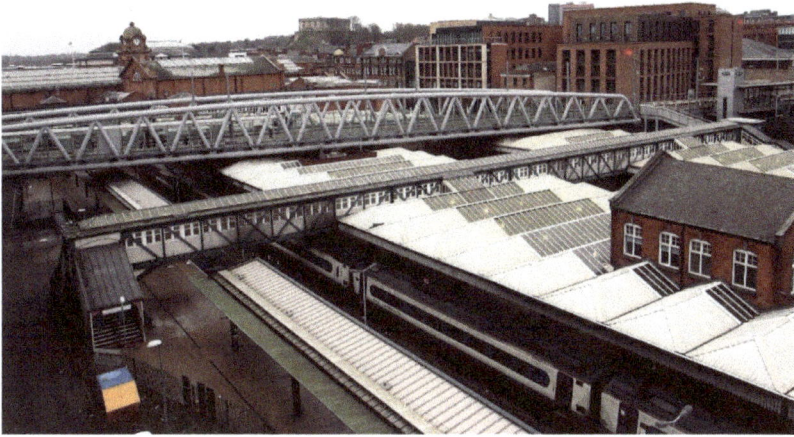

Tram Bridge in place today

Another bridge reconstruction that threw up its fair share of problems was a narrow underbridge that passed under the 4 tracks of the Derby to Leeds Line, close to the mighty Avenue Coking Plant which produced coke for industrial use from coal that came in by rail from local pits and as a by-product it also produced creosote. The bridge concerned carried the railway over a footpath that went between Grassmoor and Wingerworth and which was also used by workers going to the coking plant. As well as passing under the main lines, there was also a separate bridge over a siding and a footbridge over the River Rother, which was also maintained by British Rail. The plan was to replace the main bridge with a concrete box culvert which would be delivered to site in pre-cast sections. It was quite a small job but the difficulty was that this section of line was part of the main northeast to southwest cross-country route and there was no practical diversion available. It carried a variety of traffic throughout the weekend, including the normally "dead" period in the early hours of Sunday morning. This traffic comprised of Newspaper trains, postal trains, sleeping car trains and motorail services that carried both cars and their passengers. We were left in no uncertainty that all these trains had to run although some minor delays would be tolerated.

It was agreed that we would not start work until the early hours of Sunday morning until after the passage of several key overnight services and that we would complete the work by quite early on Sunday afternoon when daytime traffic was becoming heavier, however, that still left about 10 trains to pass during the possession itself. I didn't want to do the work over 2 weekends which would have made the job more difficult and more expensive so I had a look at the one span that we were not affecting and found that it carried a long siding which was connected to the Up Goods line to the south of the bridge and ran parallel to the main lines until it terminated at a set of buffer stops which were several hundred yards north of the bridge. I thought "if only we could take trains over the siding line, it would solve the problem" and at first, I considered installing a temporary connection to the up goods, north of the bridge. However, that would have been expensive and would have required an additional possession. Then I hit on an idea. If we simply connected the goods line directly to the siding by simply slewing the siding into the goods line and making an end-to-end connection, then trains could run past the site in both directions under the control of an on-site operations manager. Furthermore, since this operation could be carried out without affecting the main lines, it could be carried out before the main possession started – GENIUS.

I am pleased to say that all the lifts were to be carried out by the Toton breakdown crane and on this occasion, there was no real practical alternative, so I didn't have to revert to a subterfuge to keep the dreaded road cranes away from the site. The possession was short enough for me to cover the whole job, so I went up late on the Saturday evening to watch the P.-Way staff slew the ends of the siding and the up goods together and make the connection and then sat back on a beautiful summer evening to wait for us to get possession of the main lines. I had a couple of hours to while away which was passed by watching an amazing array of late-night services passing the site. There were travelling post offices in each direction with their lights on as postal staff on the train sorted the mail; there were late evening passenger services from London to Sheffield and cross-country services heading to and from the Birmingham area. There were also north bound newspaper trains and 2 or 3 Motorail services heading to and from West Country holiday destinations and a long sleeper train. I think it was the last time that I saw such a busy railway on a Saturday evening because most of those services have now fallen by the wayside. There are no Motorail services, no sleeper services on the Midland lines and both newspaper and most of the postal traffic has been lost to road and air.

Once all that traffic had passed, our movements inspector for the night arrived to take charge of operations over the siding line which was to include some single line working operations so that both Up and Down trains could be passed. Either our operations department considered this to be a most difficult operation, or the boss fancied a night shift because it was Jim Ross who turned up. Jim was the senior operations manager for the division and known to be a stickler for the rulebook. I suppose he had to be really because he set the standard for all his staff which included hundreds of signalmen spread over the whole division. Jim was a pleasant man and his daughter worked in our section in Furlong House, but I remember thinking, there will be no short cuts taken on this job. Jim's first job was to take control of the ground frame to set the points for the siding. A ground frame is just a set of points levers alongside the line which control local train movements. In order to use them, Jim first had to obtain a release form Trent Power box and then he should have been able to set the points to the siding. However, when he tried to move the levers, they would not budge. Knowing that lever frames can be difficult, I had arranged for it to be tested during the week before and it was working correctly then, so I was wondering what the problem was when Jim called me over. He had realised the difficulty, because we had already connected the siding to the Up Goods line, the track circuits had been broken and the safety locks on the ground frame could not be released. Jim simply said to me, "Brian, can you find me an adjustable spanner please?" I soon returned with one that I borrowed from our steelwork supervisor and then I watched in awe as Jim, "Mr Rulebook" Ross, calmly dismantled the safety locks on the ground frame and pulled over the points. He went right up in my estimation because a less senior man would certainly have hesitated and would have probably sent for the "Sick and Tired", which was the rather unkind name we gave to the Signal and Telegraph staff. In the early hours of Sunday morning that would have cost us at least 2 hours of delay which we could not afford.

After that little problem, the job went very well until mid-morning on the Sunday. While the crane lifted out the old bridge sections, I grabbed a late supper in the riding van and then set to work looking after the installation of the box culvert units. First, we rolled a foundation of graded rock aggregate and then topped it off with a thin layer of sharp sand on to which the concrete boxes were lowered. Generally, this went well but there is always an awkward one and one of the units would not line up, no matter how hard we tried to ease it into position as the crane lowered it down. After the third attempt, I foolishly left my finger in place as the unit banged up against the one that was already there. It hurt and it bled, and I was soon being led away to the riding van to receive first aid treatment. Just as I left site, Alan Lodge, who was duty weekend manager and was just visiting the site was heard to say, "That's it, he's lost his finger!" Fortunately, it wasn't as bad as that and I had only lost the top few millimetres but still bear the scar in the form of a fingernail that will never grow straight. While I was being looked after, the remaining units continued to be placed and I was soon back to see that all was OK. The remaining parts of the job passed without incident, and I even got treated to Sunday lunch in the riding van while the track was re-laid. Possession was given up on time and apart from a throbbing finger, I was well pleased with the job.

There were quite a few other bridge reconstructions carried out while I was working at Nottingham, including several on the Erewash Valley line. Most of them were unexceptional but I was pleased with Bridge 44 which was a single span overbridge carrying a minor road over the railway which was 4-track at the time. The original bridge had stone piers, abutments and parapets so I decided that any new work should be built to match. I felt that I could justify the extra expense because there were parapet walls on the approach to the bridge that were to remain and seeing new brickwork butting up to old stonework would have looked awful. As was usual, although I got agreement for the parapets to be built in stone, I got no extra money for the job, so we just had to do the work within the original budget. Road bridge parapets are constructed with a "core" of reinforced concrete to give them the necessary strength to withstand being hit by a road vehicle without toppling over so rather than use large stones, I contacted a local quarry who were producing stones that were cut to 100mm bed for use on houses being built in the Peak District National Park. These were ideal and not much more expensive that the ugly "artificial stone" option. I rather liked the appearance that we achieved, and I managed to scrounge a few left-over stones (firewood orders again) for my own garden wall. During a visit to the site in 2020, I found that the stones on the insides of the parapet had darkened to be a good match to the original, but the outside faces still needed to age more to blend in and had been stained from polluted air in this very industrial area.

Bridge 44 Erewash Valley – Immediately following reconstruction

Bridge 44 Erewash Valley Line – South Parapet.

Bridge 44 – Matching the stonework, Original on the left.

Bridge 44 – Parapets

My garden wall

The final bridge reconstruction that I will tell you about is Bridge 34 on the South Leicester Line at Blaby. It carries the railway, which is 2-track here, over the Leicester arm of the Grand Union Canal. The railway crosses the canal at an angle of 30 degrees which makes the bridge longer than it otherwise would be. There are 2 ways to achieve a crossing like this; one is to build abutments that are directly opposite each other thus creating the shortest span for each beam but building a bridge that is longer than necessary and which has triangles of unused area above the beams. The other, more common, way is to build abutments that are offset from each other and then span between them with longer beams that are parallel with the tracks. This creates the smallest bridge area but needs stronger beams. Bridge 34 was built using the former system and had a series of wrought iron girders spanning between abutments that supported brick jack arches. Jack arches are brick arches that span between girders, normally springing off the bottom flanges. They are a very strong type of construction, and the brickwork also protects the main girders from corrosion, but the exposed bottom flanges do corrode, and this leads to suspicion about the general condition of the girder webs and top flanges which are hidden. Usually, when investigated by taking out small areas of brickwork, the webs prove to be in perfect condition but hidden features like this are always difficult to inspect, so when bottom flanges are corroding badly, the decision, as in this case, may be taken to reconstruct the bridge rather than repair. Demolishing a bridge of this size can be a time-consuming job because the jack arches must be broken out before the girders can be lifted off. However, as usual, time was not in abundance and we were allowed just 30 hours line closure for the work which started after the last Saturday night service and had to be finished before the first passenger train on Monday morning. My colleague, John Colebourne, had already reconstructed the sister bridge No. 33 a few months before but this was over a public road so he could use heavy plant and road cranes to assist him and lorries to take away the large quantities of broken bricks from the demolished jack arches.

The first problem that we had to overcome was how to break out the jack arches and remove the rubble. This had to be done as quickly as possible and all our initial ideas were clearly going to take too long. The only quick way was to use excavators working from track level (after the removal of rails and sleepers), which were fitted with hydraulic breakers to hammer out the brickwork and let it drop down to the canal below. However, the British Waterways Board would not have thanked us for filling in the canal so we hit on the idea of bringing in a canal barge which would sit under the bridge and into which the rubble would drop. Once full, it was intended that the barge would sail down the canal a few hundred yards to a point where it could be unloaded and the bricks stored for removal later in the week, the barge would then return to site for more loads. We also got hold of a small digger that could stand on the towpath and load any bricks from there into the boat. Seemed like a good plan to us and we got approval from the BWB, so the next step was to find a boat. This section of the Grand Union is a wide canal from Leicester to Foxton Locks, so we needed a wide boat with a large open hopper for the rubble. Not only would it have much more capacity than a narrow boat but it would also occupy the full width of the canal, leaving only small gaps for rubble to fall through into the water.

I was confident of the plan but getting hold of a boat proved difficult. I had a contact at a boatyard in Leicester with a man who had done some work for us before, providing workboats for bridge repairs, and although he had nothing big enough for this job, he did help in the search, and we eventually found a suitable craft available at Goole on the Humber. This was quite some distance away, but the owner of the boat was quite confident that he could do the job, so we gave him a contract for the work. As the reconstruction weekend approached, we found it difficult to contact him (no mobile phones in those days) so we just had to assume that he was on his way. His route would take him down the Humber to Trent Falls where he would enter the River Trent and continue on through Gainsborough to Cromwell lock, just north of Newark-on-Trent. All that part of the journey would be in tidal waters so he would have planned his journey to enter the River Trent after low tide and then travel fast, with the tide behind him to Cromwell. After that, the journey along the Trent to Nottingham would not be difficult, with all locks being massive electrically operated structures designed for freight traffic. At Nottingham, he would pass through the city along the Beeston Canal with hand operated locks before re-joining the Trent at Beeston. From there onwards, the route ran to Trent Lock where he turned south along the River Soar navigation to Leicester and eventually to the site at Blaby. There was still some gravel boats operating north of Leicester at the time so I was reasonably confident that the boat would get as far as that but the final section to Blaby was used mainly by narrow pleasure craft, so getting a big barge along the final section might prove difficult.

By Thursday, there was still no sign of the boat so I contacted the British Waterways Board and after a few phone calls, they were able to tell me that a mysterious large boat had been heard travelling up the Trent by the lockkeepers in the middle of the night, but they couldn't confirm if it was my boat. I just had to cross my fingers and wait until Friday afternoon when I went down to Leicester and found the boat safe and sound in the boatyard. I was mightily relieved but the journey to site could not be completed until the time of the agreed canal closure because once at the bridge, the boat would effectively block the canal, so it was Saturday afternoon before I was finally sure it had managed to get to site. It was a lovely boat, smartly painted and fitted with an end cabin in which its owner ate and slept. It was clearly his pride and joy so I was a little worried about the damage we might cause it we accidently dumped some bricks on his cabin roof.

John had used a large road crane to reconstruct bridge 33 but that option was not available to me because there was no road access to my site, so it was back to the old standby, the rail breakdown crane. The problem was that because of the skew of the bridge, it was difficult to get the crane close enough to the bridge to do all of the lifts. For a while it looked as if I might need 2 cranes to work in tandem, but I worked on the problem, eventually producing a lifting plan showing exactly where the crane would have to be positioned if the job was to be done safely.

I took Chris Shardlow, the crane supervisor from Toton to site and we marked everything out on the ground and decided that it could be done – just. Of course, the track and ballast had to be removed before work could start but once that was done, we moved the crane into position which had to be right up to where the track had been cut. The crane lifted off its own match wagon and relieving bogie and put them on the side of the line, it then approached the end of the line very carefully under its own power with the supervisor walking along with his hand on the brake valve so that the crane could be stopped just before it toppled off the track. Once that was done, the outriggers were out in place, and everything screwed down very tightly before the crane was ready to start work. While all this was going on, we had two excavators equipped with hydraulic breakers on the ends of their digging arms and these were busily breaking through the jack arches and allowing the brickwork to fall into the barge below. Once the barge looked full, I stopped the diggers and roused the boatman to take the rubble to the unloading point. He was in his cabin watching television and, incredibly, he had failed to notice that the barge was well and truly stuck on the bottom of the canal!

Canals today are not as deep as they were when commercial traffic regularly passed by, clearing the silt. We had to unload about half of the rubble before the barge could move. This was done by the small digger that we had standing on the towpath. The rubble still in the barge was then taken away and unloaded; it then returned to site and the rest of the rubble was re-loaded and taken away. After all that, there were still several more loads to be taken away before all the brickwork from the jack arches had been removed. This caused a bit of a delay and by the time all the old girders had been removed, we were more than 2 hours behind schedule. However, the rest of the job went well and by Sunday afternoon, the beams had all been placed and the job of joining all the beams together with post-tensioned cables was taking place. This involved threading cables through holes in the beams and then tightening them together with hydraulic jacks to form a sold concrete deck rather than a series of separate beams. Once this was done the ducts through which the cables passed had to be filled with cement grout, but I had forgotten to organise a supply of water! The only solution was to use water from the canal for the job and we carefully lowered a pump fitted with a filter down into a clear part of the canal. I felt that this would be OK because we had avoided letting any obvious vegetable matter into the water but of course, water that is not from a mains source should have been sampled and tested. Unfortunately, while we were doing all this, Keith Ratcliffe, the AACE (Works) turned up on site with a senior engineer from head office who had come to look at progress. They didn't stop what I was doing as both knew that to do so would result in and over-run, but Keith took me to one side and told me of his displeasure. He also instructed me to get the water sampled for testing after the event although what we would have done if it proved unacceptable, I have no idea.

The rest of the job went well, and we finished the job on time. In fact, I can honestly say that I never had an over-run on a bridge reconstruction although sometimes it was a very near thing!

Bridge 34, Blaby over the Leicester arm of the Grand Union Canal

Chapter 14

Return to Peterborough

In 1989 my search for a new job bore fruit when I spotted a post being advertised back in Peterborough that looked as if it was tailor-made for me. It was for a Works Construction Assistant at the new area office in Peterborough, back in my hometown. The area had recently been created to be one of only 2 that would maintain the East Coast Main Line between King's Cross and the Scottish border and was a part of the process known as "sectorisation", which would eventually lead to "Inter-City" becoming not just a brand name for British Rail's express services but a business in its own right. I reasoned that, if I got the job, there would be no need for me to move house immediately because the journey time to Peterborough was less than an hour and if, eventually, I wanted to move at least it would be to somewhere that I knew. It also fitted in with family commitments because my mother still lived in Peterborough, and I would be able to make her a weekly visit.

The job itself was essentially the same as the one that I was already doing at Nottingham but with the added responsibility of having control of the transport fleet which meant it was one grade higher. I applied and was duly called to an interview in York with the Chief Civil Engineer (CCE), Eastern Region and my prospective boss at Peterborough, Alex Turner. The interview got off to a shaky start because Brian Davis, the CCE, asked me about my career and when I got to the point where I left the Eastern Region to go to the Midland, he exploded! "Why the hell did you leave the best railway in the land to go to a tin-pot organisation like the Midland," were more or less his exact words and all I could do was mumble about "promotion" and "broadening my horizons", but I did manage to mollify him somewhat by saying that I always wanted to come back to my roots. I also noticed that he kept asking Alex questions at the same time and only later did I realise that Alex was effectively being "interviewed" too because he had not yet been permanently appointed to the job and Mr Davis was still to decide.

The rest of the interview seemed OK, but I didn't think that I had done particularly well. I had mixed feelings about that as I travelled home by train. There were severe delays on the journey, and I began to wonder if commuting to Peterborough by two trains each-way every day would prove to be a pain. As it transpires, I don't think that I had too much opposition to contend with and I was duly appointed to the job. Arrangements were made for my early release from my present post (maybe they wanted rid of me) and I found myself on the train to Peterborough on a Monday morning about 2 weeks later. The journey proved to be no problem and when duties demanded a different start or finish time which didn't fit in with my local trains, it proved easy to drive to Grantham and leave my car in the station car park with a railway Hi-Vi vest over the steering wheel to ensure that I didn't pay car park charges!

The Peterborough Development Corporation had for many years been promoting the town as a good place to live and work with the slogan "The Peterborough Effect" and this was exactly how I felt working there. There was a tremendous "can-do" attitude amongst the staff which I found incredibly refreshing and new ideas were welcomed and encouraged. There were probably many reasons for this including the good quality of the staff, the leadership of Alex Turner and the idea that we were now working for a flourishing and expanding business rather than a shrinking and, by some, despised British Rail. There were other factors too, many of the staff that worked for us and with us had been involved in the electrification of the East Coast Main Line, which had been a very successful project and one undertaken with a great deal of gusto. It came in on time and under-budget; perhaps today's engineers could learn a thing or two from that scheme although the practise of working on the wires under lookout protection and staying hanging on to the wires while trains passed below would probably not be acceptable today. Whatever it was, I thoroughly enjoyed my time there in all three of the posts that I occupied.

Works construction assistant was a wide-ranging job which involved planning all our work and attending the weekly planning meeting when competing demands for possession time and isolations were resolved. This meeting involved operating staff, P-Way, Works, S&T and Overhead line. It could become a little fraught at times, but I enjoyed the cut and thrust. The other parts of my department such as Works Study and Transport could, in the main, be delegated to my assistants. My main thrust was to make sure that the physical work being carried out by our staff from all the depots was being done as well as possible and to do that I had to get to know the people at the works depots of which there were 4.

Palace Gates depot covered King's Cross Station itself and the East Coast Main Line as far as Wood Green plus the lines to Moorgate and Dalston Junction. Nobby Clarke was the supervisor with assistance from Brian Wyant, George Lovell, Peter Colgan and the chargehand painter Mr Roberts. I never knew the forename of the latter; he was always addressed as Mr Roberts, even by Alex. Our painting gang at Palace Gates was entirely made up of men with an Afro-Caribbean background and Mr Roberts was not only their chargehand but also their spokesman who would look after all their interests. He was an excellent man, and he ran a good team of painters who were often involved in working in the public areas at King's Cross where a top-quality job was always expected. When he retired, he kindly invited me to his retirement party at the King's Cross Railway Social Club which I thoroughly enjoyed. Caribbean food was provided with plenty of drink and a good time was had by all. Mr Roberts was intending to return to his roots on a small Caribbean Island and I had a standing invite to visit him there, sadly I never did make it. As well as all the usual trades, Palace Gates boasted a French Polisher. His post stemmed from the days when boardroom furniture and special waiting rooms for the rich and famous were the order of the day at King's Cross. There was little of that sort of work for him, so he was mainly used as a carpenter. Bearing in mind the other opportunities for tradesmen in the London area, I had rather expected the Palace Gates staff to be of rather poor quality but, in fact, they proved to be some of the best that I have encountered on the railway.

I make no apologies for naming many of the depot staff because they helped make my life at Peterborough a real joy. Of course, it wouldn't be a railway depot without its fair share of characters and incidents and most of the ones at Palace Gates seemed to involve Brian Wyant. Brian was a good supervisor but somewhat headstrong and, it seemed, unlucky. We had a job at the front of King's Cross Station in an area where the local undesirables gathered to drink and generally annoy the public. They sat on a low wall which surrounded a little-used area so in conjunction with the station manager we decided to build a new wall, topped with railings that would exclude access to the area and remove any opportunity to sit down. Within the railed off area, we planned to plant a substantial tree. This was a real old-fashioned railway job, and we ordered a tree from Poppleton Nurseries, near York and had the steel railing from Woodburn Shops in Sheffield. Both establishments were a part of the BR family, the nurseries at Poppleton having been established by the North Eastern Railway back in the 19th century. It continued to supply plants for stations and other BR establishments until privatisation when it fell into rack and ruin. It is interesting to note, however, that a group of ex-railwaymen now run it as a charity. Brian was in charge and to plant the tree, he organised an excavator fitted with a breaker to remove the concrete paving. Unfortunately, in doing so, he broke into a London Underground passage below. Fortunately, it was only a ventilation and services tunnel so there was no danger to trains or the public but to say that LU were rather annoyed would be putting it mildly. On another occasion, when Inter-City organised a non-stop high-speed train from Edinburgh to King's Cross in an attempt to establish the shortest ever journey time, it was Brian Wyant in charge of a drainage job at Holloway that managed to cut a signal cable and delay the train.

We did quite a few interesting and innovative jobs from Palace Gates depot. At King's Cross Station, staff stood down at track level to "water" the trains, a task that involved attaching hoses from standpipes to the filling points on coaches so that the water carried for toilets and the like could be replenished. It was a dirty job and not without safety risks, so a scheme was developed to install watering points along the edges of the platforms. Metal boxes with hinged covers were fitted along the platform edge, which, when closed form part of the copings. They contained a water tap and short length of hose to attach to the train once it had come to a stand. The specification was that the boxes were to be as close as possible to the watering points on the train to avoid any trip hazards caused by trailing hoses and for this reason, they were spaced out to suit both Intercity 125 and 225 trains. The main part of the scheme was to install new water mains along the length of each platform to form a new supply. These mains started off at 150mm and gradually reduced to 25mm at the buffer stop ends. Some of the platforms at King's Cross were solid, so the new mains were installed in trenches, but most were hollow, so the new mains had to be installed by crawling under the platforms. It was all night-time possession work, as platforms could not be obstructed during the day.

I put Paul Lawrence in charge and sent him on a water industry course to learn to fusion weld the new "plastic" pipes. He did an excellent job, the results of which can still be seen today. There were two "flyover" bridges on the Place Gates patch, one a Ferme Park to carry trains crossing from down yard to up side and vice-versa without conflicting with mainline traffic and one at Wood Green that carried the down Hertford Line over the main lines. Both were due to be painted and, along with several others further north on the line, it was decided to impress Chris Green at NSE by using the colours of Network Southeast (Blue and Red). This proved to be a somewhat controversial choice for some locations, but it found good favour in the London Suburbs. I was determined that any painting that we did would be long-lasting and to my mind, that could only be achieved by first blast cleaning the bridges to remove all old paint and rust before applying new paint to the full BR specification. It had been the practise to use contractors to do any blast cleaning work but since we had a good programme of bridges to paint, covering 3 of our depots, we decided to do all the work ourselves. Several staff were sent on blast cleaning courses, including Peter Colgan from Palace Gates. When carrying out blast cleaning, the operative wears full protective suit and helmet so is effective cut off from the world. Blast cleaning is very difficult to do well at night for obvious reasons, so we looked at ways of doing as much as possible during the daytime. Possessions were not available, so we devised a system of lookout protection to warn of approaching trains with Peter being told to stop work and stand clear of the line by the use of a telephone line into his helmet. It worked very well but would probably not have stood up to risk assessment at the time (what happened if the telephone line failed?) and would certainly not be approved of today.

Kings Cross Station – Platform watering points

Ferme Park Flyover

Heading north, the next depot was at Hitchin, just around the back of the station. The Works Supervisor here was Bernard Dawson, assisted by Dick Worby and Colin Teague with help from several chargehands and bridge examiners that included Pat Hogan, Chris Richardson, and Dave Pile. Bernard ran the depot as his personal fiefdom, but he certainly got things done. He wasn't afraid to put me right when he thought it necessary and he wanted to know everything that was going on, not only on his own patch but also at Palace Gates depot which I think he regarded as a satellite of Hitchin (at least 2 of the Palace Gates supervisors had trained under Bernard). He was an excellent supervisor and we got on very well. Dick Worby suffered with severe tinnitus, his ears being damaged during the time that he worked as a blacksmith, Colin Teague and Pat Hogan were excellent and Chris Richardson, who was a chargehand with a background as a steel worker, could be relied on to undertake any project. Dave Pile was a rough and ready Bridge Examiner and bricklayer who could be relied upon to look after major projects and who followed me later into the private sector, as did Bernard Dawson himself.

The depot looked after the East Coast Main Line from Wood Green to St Neots and the Cambridge Line from Hitchen to Royston. In all, there were some 25 Network Southeast Stations, 7 tunnels and several viaducts, including Welwyn Viaduct. We did quite a lot of repairs to the latter while I was there and somewhere on the structure, out of reach of vandals, is a plaque dedicated to an apprentice bricklayer who worked on it and sadly died young. The tale of the fire at New Barnet Station is given a separate chapter and epitomises the spirit at Hitchin Depot and was only one of the jobs that we carried out at Network Southeast Stations. Chris Green was in charge at NSE when I first got to Peterborough (he later became managing director of InterCity) and he had a wealth of ideas for improving the image of his railway, one of which was to paint all the station lamp-posts red. On other parts of NSE, this work was given to contractors without any input from the respective engineers, but we were not about to allow that to happen on the GN.

We approached NSE and told them that we wanted the work and they gladly agreed on condition that it was carried out within a very short timescale. To do this we put every painter that we had onto the job from the three southern depots and even brought in a team from Doncaster. On top of this, Bernard selected the best of his labourers to support the painters as "Brush Hands". The work was all done on time and contributed to all staff on the NSE Great Northern lines being given medals to celebrate the outstanding performance in 1989. I still have mine today! Working for NSE was always interesting and sometimes exciting. It was good to embrace the Chris Green "can do" approach and to support him in any way that we could. When he wanted station improvements, most of which were on Bernard's patch, we were only too pleased to assist and one of the projects, led principally by Chris Richardson, was the relaying of many of the station platforms in "block paving", which was a fairly new system at the time. However, it was my idea to go one step further and incorporate the NSE logo into the paving pattern and some of these are still clear today. Bernard's team with Dave Pile in charge as acting supervisor also extended several platforms on his patch to accommodate 12 coach trains and that work kept us busy for many months.

Of course, nothing is all sweetness and light and one incident, which seems funny now, resulted in serious charges being levelled against one of Bernard's men and considerable embarrassment to my department. I got in to work one morning to be met with a message that there had been a major bomb scare during the night which had severely delayed services for several hours and that one of my staff from Hitchin depot had been detained by the police. It transpired that one of Bernard's men who was known to be a bit of a joker had decided to pull a stunt but since it was a time of high alert because of IRA bomb threats, he made a very unwise decision. Piecing together an old battery, some sticks of epoxy resin that looked for all the world like gelignite and a few electrical wires, he fashioned a mock bomb and put it under the lorry at Bernard's depot. His intention was to give Dum-Dum the lorry driver a fright when he found it the next morning but unfortunately for him, the lorry was due to go out on a job that night and when Chris Richardson arrived at the depot to pick it up, he spotted the device and in the dark, he thought it might be a real bomb and called the police. Since the depot was close to the line, they also informed control who stopped all trains and called a major alert. The police arrived, followed by the bomb disposal unit and they refused to believe the protestations of Dave Pile, who said that he could see it wasn't a real bomb. After several hours of mayhem, the device was declared a fake and the lines were re-opened before the morning passenger rush, but the culprit was arrested, imprisoned and sent for trial on serious charges.

Welwyn Viaduct with scaffold in place

Block paving platform with inset NSE logo at Welwyn North

Peterborough depot covered a long section of the ECML from Huntingdon to Newark-on-Trent, but it was not a very happy place when I first arrived. The old supervisor, Dick Lowes, whom I had known from my first time at Peterborough, had retired and his assistant was in temporary charge. One of the first tasks Alex gave me to do was to find a replacement for Dick and he made it quite clear to me that it was not to be his deputy. Unfortunately, the man himself thought that he would automatically get the job and was behaving as such. There were many good men at the depot, but I realised that if I promoted one of them over his head it would create a difficult situation, so I looked around for a possible replacement. Of course, the post had to be advertised on the railway vacancy list but when it was, I contacted Paul Moffat who I had previously appointed to the post of assistant supervisor at Kettering where he had proved to be a great success. Paul lived in Corby, which was a shortish drive to Peterborough, so he would not have to move home and he agreed to apply for the job and was duly appointed by Alex, after he had interviewed him. Paul found his early days at Peterborough difficult and at one time he was close to quitting, but he eventually settled down and turned the depot into one of our best. Pat (the former deputy) stayed on as Paul's assistant alongside Russ Pettit and Mick Chambers.

There was plenty of routine bridge work being done at the time as well as major brickwork repairs to Stoke Tunnel and Peascliffe Tunnel which I have described elsewhere but some of the biggest jobs tackled by Paul and his team were station and building works. Our own office at Peterborough, called Grant House had been internally fitted out to an acceptable standard but in the rush to move in, there had not been time to attend to the external works and the defective roof, in particular. This was originally a slated roof but in typical railway fashion, when it had started to leak at some time in the past it had been treated to the appalling process of "Turnerising". This involved coating the whole of the roof with a bituminous solution which certainly stopped any immediate leakage but totally ruined the appearance of the slates and effectively prevented proper maintenance of the roof in future. Because of this, tarpaulins had been spread over the slates as a temporary solution but on windy days, I would receive a call from the solicitor's office opposite to say that our tarpaulins were coming loose again. Most embarrassing for a civil engineering department, I have to say. There was some money in the budget but nowhere near enough to pay for the complete re-slating by a specialist roofing company. However, Mick Chambers was a roofer by trade and expressed a willingness to undertake the work. So, using a team selected from Peterborough depot, he completely stripped the roof (a filthy job) and renewed all the slates. It looked great when he had finished and financially, it had not only been cheaper than using a contractor but provided good employment for the depot staff during a time of tight budgets.

Peterborough depot carried out works at Grantham and Newark Northgate Stations in particular, responding to the requirements of InterCity who wanted good quality buildings for all their stations and also needed platforms to be extended to accommodate the longest of their passenger trains. Platform works are always a challenge, particularly against main lines but Paul Moffat already had experience of this sort of work from his Midland days and he made sure that a good job was done. Not only was platform 4 extended at Grantham but large parts of platforms 1, 2 and 3 were re-built and raised. The main roof and awnings on Platform 1 at Grantham were also refurbished and made waterproof. The opportunity was taken to remove layers of paint from the metal parts of the structures, including columns. In the end, they looked much better for it, but we made a real "rod for our own backs" because the blast cleaning also removed all the filling from blowholes in the cast iron of the columns leaving them looking like honeycombs. Again, it was an opportunity for our staff to prove that they could not only do good work in tunnels and other areas where strength was the main concern but also in public areas where appearance and craftsmanship were required.

They responded with alacrity, and I was really proud to show off our work to a series of InterCity line managers. A similar job was done on the awnings of platform 1 at Newark with similar excellent results. Paul Moffat and his crew were determined never to let the side down and sometimes took this to extremes. There was an ongoing programme of providing a safer cess path for track workers, particularly patrol men and at Essendine this meant constructing a low wall of stone-filled gabion baskets at the toe of the cutting slope. Of course, this was weekend work and Paul and his crew duly turned up on a Sunday morning with men and excavators to carry out the work. The Gabion baskets were there waiting for them but no stones to put in them! "Someone" had forgotten to order them (I heard the name of Pat O'Boyle mentioned). Paul was not amused and after a (long) rant over the phone to the person responsible, he set about saving the situation. Gabions are designed to be filled with rocks or any other heavy material that will create a block unit to form part of a retaining wall and Paul reasoned that bricks would do the job. He looked around and in Essendine yard there was a disused weighbridge cabin, so Paul got it knocked down and used the bricks to fill gabions, thus "saving" the job. I saw them when I visited the site during Sunday morning "rounds", and though they didn't look too good, they did the job so I was happy. Much less happy was the owner of the cabin who turned up on Monday morning to find it had been demolished! It transpired that that part of the old station yard had been sold by BR and the sale included the cabin. I had to fend the call as best I could and told Paul to go and see him and sweeten him in some way, which he duly did. I didn't ask how but I know that those sorts of transactions usually involved something like a lorry load of old sleepers. In the run up to privatisation, Peterborough depot was at the forefront of our efforts to win contracts for work beyond our area and this is covered in a separate chapter.

Bridge 144, just south of Huntingdon on the East Coast Main Line was on Paul's patch. It is a significant bridge which carries the 4-track railway over the River Ouse on three metal spans. Up side and down side are of different vintages with the up side being the more recent. The down fast spans are of the way-beam and longitudinal timber type and as far as I am aware, they are the only longitudinal timber spans in the country that carry 125 MPH track. Whilst I was there Paul and his men renewed several of these timbers, but I left this work in the capable hands of Eric Jeyes, the bridge assistant. Longitudinal timbers are never without their problems, but we managed to always maintain them without the need for temporary speed restrictions. There was also a major scour protection scheme done at this bridge which I cover in chapter 17, but by far the most controversial work was the simple painting of the bridge.

Following my own inclination to support Network South-East as far as possible by using their house colours when we re-painted a bridge and with the full support of Alex Turner, we first did a full blast clean of the structure and then coated it with BR spec. protective paint in a colour scheme of red girders and stiffeners picked out in Blue. I believe the NSE management were well pleased with the result, but this bridge was not located in a north London suburb but in rural Cambridgeshire where our big red bridge was not to the liking of many of the local landed classes who preferred that the railway should pass through their county without being noticed. At about that time Alex Turner was undertaking his annual inspection of the line in the "Engineers Saloon" and it was his habit to take the day's incoming letters with him to open while the "Saloon" travelled from site to site. He opened one particularly vociferous complaint about the red bridge and so he asked the 4 other people on the saloon what their views were. All said that they liked the bright red colour, so Alex immediately penned a reply to the complainant saying that he had recently carried out a survey and the response was a 100% approval!

Grant House, Peterborough

Extension to Platform 4 at Grantham while an InterCity 125 passes through Platform 1.

Bridge 144 – The controversial painting scheme.

Doncaster depot had a large area to look after, stretching from south of Retford through Doncaster to Heck on the East Coast Main Line and most of the line to Leeds via Wakefield Westgate. The Doncaster area itself provided plenty of work with a major station and other buildings to look after. Ron Filby was in charge when I first arrived, supported by John Darton and Graham Marris. Ron Filby eventually retired, and the extremely reliable John Darton took his place. The depot had our only steel gang and they found themselves being sent up and down the line and helped us undertake several projects that would otherwise have been beyond our capabilities. At Welwyn Garden City, they worked under the supervision of Chris Richardson from Hitchin Depot to refurbish the footbridge that connected the station to the new Howard shopping centre and install new lift shafts and a link new span to connect the station directly into the shops. Outside of the routine works, Doncaster, like the other depots undertook a lot of interesting station works and they became particularly good at the improvements of platforms by restoring them to the correct height and installing new platform copings.

We had many station projects that involved replacing the platform copings for various reasons and at first, we struggled to find the best way of doing the job. The traditional way was to lift the copings off by hand, using wedges or "pins and feathers" to first break the mortar bed. "Pins and feathers" are just a wedge system fitted to a compressed air breaker. New copings would be placed on a mortar bed, again by hand or possibly hanging on chains from some sort of lifting device but because the chains had to be slung around the copings, they had to be eased up to release the chains, thus disturbing the carefully prepared bed. Overall, it was a slow job and one that could result in back-strains and trapped fingers and often only 10 new copings might be placed on a shift. We had to do much better and I set out to find a better way.

The first piece of equipment to offer substantial improvement was a suction lifter. This was a simple device that could be placed on the top of a coping attached to it by suction so that it could be lifted by crane. Then came the choice of lifting machinery, which in the end boiled down to either a "Bobcat" miniature excavator operating along the platform or a road/rail excavator operating along the adjacent line. The "Bobcat" had the advantage that work could be started as soon as possession of the line was granted because it did not need isolation of the electric wires to be able to work. However, it was a slower operation as the "Bobcat" had to manoeuvre along the platform and around obstacles. The use of a Road-Rail excavator with suction pads was undoubtedly the fastest operation and we had a good chance to try it out on platform 4 at Retford which is on the low-level line from Sheffield to Lincoln and was not electrified. I knew that John Darton intended to "give a real go" but even I was surprised when I turned up on site to see half of the platform had had its copings removed and John was just starting on laying the new ones. This would mean that we would replace all the copings on half the length of the platform, some 100m, in one shift. I was apprehensive and feared that when the first Sunday afternoon train arrived, there would be nowhere for the passengers to go but all went well, and I think we must have set some sort of record.

We used a road-rail machine called a "Mecalec", which has a unique three-part boom enabling it to manoeuvre loads more accurately within confined areas and these proved ideal for the job. The problem was that when we tried to use this machine "under the wires", we had to wait for an isolation before we could start work. Our electrification colleagues in the OLE department had limited resources to provide isolation on a Saturday night and had to give priority to major works such as track relaying so it could sometimes be 3 hours after getting possession that we were able to start work; clearly this was not satisfactory, so I started to look at an alternative. In conjunction with the "Mecalec" company, we fitted a machine with electronic and physical "stops" so that it could operate under live wires. This limited the machine's capacity, but it was still able to do the platform work. It was also fitted with additional earthing equipment and put through stringent tests by the OLE engineers before being approved for work. The rules were that the machine had to be put on the rails away from the OLE, usually in a siding, and then travel to site in its "locked-down" mode. It proved a great success, and the company modified several machines to meet our requirements. The biggest job that Doncaster undertook was to build a new station at Bentley in 1992. This was carried out for the South Yorkshire Passenger Transport Executive and comprised 2 platforms and waiting shelters. Not a large station but projects like this gave everyone a fillip, particularly at a time when the treat of privatisation was beginning to raise its head.

Bentley Station
By Gregory Deryckère - http://www.vlaki.com, CC BY 2.5,
https://commons.wikimedia.org/w/index.php?curid=628028

I have mentioned many of the staff at the four depots that helped to make it a happy time for me, but I shouldn't forget the many office-based staff who were equally good. There are far too many to list here, but I won't forget the way that they made me welcome. They included the Jeyes family – Father Terry, whom I knew from my first days on the railway and now joined by sons Eric (in charge of the bridge section), David, Phil and probably several others. One problem that did raise its head was a lack of transport for technical staff. Basically, we were one vehicle short of having enough to allocate one to each section and it always seemed to be Eric's bridge section that missed out. I set out to resolve the problem and got to hear that there was a spare long-wheel based land rover at York that was no longer wanted by the soil mechanics section. I welcomed it with open arms and presented it to Eric, Mark and the others in the bridge section thinking it would make their friend for life. Ungrateful lot, all I got was complaints about the petrol consumption (16 mpg) and the lack of power steering. I had to agree it was a very uncomfortable vehicle, but it looked good on a railway site! I spent three great years as Works Construction assistant but that was not the end of my time at Peterborough and my next promotion came as a complete surprise to me.

Chapter 15

Permanent Way Stuff and on-call at

Peterborough

Within a couple of weeks of my arrival in Peterborough, I was put on the roster of duty managers which meant being on-call for 24 hours a day for one week in three to deal with any emergencies that arose (both works and permanent way). I was also required to visit all the main engineering sites over the weekend, including major renewals work. It sounds onerous but in fact it was quite fun although there was very little extra money involved. My first weekend, I went out on the Saturday afternoon to make myself acquainted with the staff on a big relaying site just north of Peterborough. I had arranged to meet up with the technical assistant in charge, Don Ryan, who was known to be a real character and who had already incurred the displeasure of Alex Turner by using his large office table for "recreational purposes" (I must leave the reader to guess the exact nature of the recreation). However, I found Don great and once he realised that I was there more to learn than to dictate or criticise, he gave me the grand tour of the site, including a spell in the cab of a tamper. There was a third job incumbent upon the duty engineer, and it wasn't long before I found myself out on the track on a Sunday morning with Alex as he was inspecting the site of an overnight relay in order to approve the works and agree that the line could be opened at full line speed, an operation that was referred to as "handing back". This means that the line, which had been in the "possession" of the engineering department, would be "handed back" to the jurisdiction of the operating department who could then begin to run trains.

The East Coast Main Line at the time was unique in two aspects. Firstly, much of the route was officially available for 140 MPH running. In practice, although track, signalling and trains were all ready for 140 MPH, it was never implemented for regular passenger services because various issues associated with the safety of staff on the line and the public at level crossings were never resolved but frequent "specials" were run at that speed, both for testing and publicity purposes. The second unique feature was that we were not allowed the use of speed restrictions following weekend engineering work after 6 o'clock on a Monday morning. Instead, we made full use of a "yellow machine" called a "Dynamic Track Stabiliser" or DTS for short which worked along with the tampers to produce new track that was not only correct for alignment and level, but which was consolidated so that no deterioration would occur once the line was open for traffic. The DTS provided a graph known as a "TRACE" for the Handback Engineer to look at and which showed the condition of the finished track but that was only the first stage of the procedure. The engineer would also be apprised of the work by the technical staff involved and they would let him know if any problems had occurred that might affect the finished condition of the track. After that, the engineer would physically walk the track, usually with the relaying supervisor at his side and check that all fastenings were in place, there were no obstructions such as tools left on the line, that the correct ballast profile had been achieved, and that all welds had been made correctly.

As the engineer walked the track, he would check the cross-level every 3m and record it, comparing it with the designed cross-level and ensuring that there were no "twist" faults. The latter occurs when one rail is lower or higher than it should be, producing a "twist" in the track which will cause a train to rock and in extreme circumstances, to derail. There were strict limits on the amount of twist fault that could be allowed if the line was to be re-opened and if this was exceeded, the engineer would either not allow the line to re-open or allow opening at a lower speed. Finally, the engineer would take a good hard look at the overall track alignment although, since many handbacks were done at night, this could be of limited use. Having followed Alex around on one site, that was the extent of my training, and I was officially one of the area's handback team which, because there were usually several relays each weekend, meant that I was out most weekends doing the job.

On-call duties themselves were not normally onerous and most calls could be dealt with over the telephone but occasionally there was a need to go to site to sort out a problem. One of those was a major fire and I have given that a chapter to itself but there were several others, including one that was almost the undoing of me. At the time, I was staying one night each week with my mother, who was elderly and living alone in the old family house in Fletton. She would cook me tea and we would spend an evening reminiscing before turning in. One night when I was sleeping there, I was awakened by my pager going off in the early hours of the morning. When I contacted control, they told me that there had been a de-wirement on the Fens south of Peterborough. This meant that the pantograph of an electric train had brought down a length of overhead wires. I queried why I was being asked to go, because dealing with the problem was the responsibility of the OHLE staff and I was told that the OHL engineer thought that the track had moved, causing the pantograph to run off the wires. As any P-Way engineer will tell you, that is as likely as seeing the Pope's arse, as a good friend of mine would say but I had to attend because of the peculiarity of the line as it crosses the "Fens". The ground of the fens is soft and wet and rather unstable, when the line was built, wooden faggots and sheepskins were thrown into the mire to prevent the new track sinking out of sight. To this day, there is a speed restriction of 100 MPH on all trains and the line is tamped regularly to correct the defects that appear over time. Therefore, it was (just) plausible that the condition of the track had been a factor in the de-wirement.

Anyway, I got out of bed, put on my site clothes and drove the 5 miles or so to the access point, somewhat congratulating myself on being able to respond so quickly by virtue of the fact that I was already staying in Fletton. It was, of course dark and I followed the time-honoured rule of walking "facing traffic" along the line from the small farmer's crossing where I parked my car towards the lights of the OHLE staff that I could see about half a mile ahead of me. Thus, I was walking south along the down line, keeping a sharp lookout for the lights of any approaching train. What I failed to appreciate was that the incident had caused the up line to be blocked and the down line was being used by trains travelling in both directions; that was until, I looked round to see a southbound parcels train just a couple of hundred yards away and heading straight for me.

What made me look round, I don't know, maybe I heard something or felt a vibration but whatever it was, it saved my life. Even then, I might have ignored it, because I expected any train to pass me on the up line. I jumped clear and stored up the experience to remind me of my stupidity and to use as an example when, later in my career, I became involved in safety training. Once I got to site, I quickly established that the track was not at fault and even the OHLE engineer had to agree with me that the problem lay with a supporting mast which has succumbed to the problem of fen "shrinkage" and had moved out of plumb thus the wires were not in proper alignment with the train pantograph.

On another Sunday morning, I was just driving along the A1 when I was contacted on my "Band 3" radio to be informed that the Ballast Cleaner working on the slow road near St Neots had derailed. This was not a particularly unusual event, but I knew from experience that if it was not dealt with quickly, there would either be a possession over-run, or we would have to curtail the work and neither of these options was particularly attractive. I got to site at about 10am and was told that the re-railing team from Hornsey depot had been contacted by control but they were unlikely to get to site until mid-afternoon. That was no good at all, so I authorised an attempt to re-rail the cleaner with our own resources. The derailment had occurred when the machine rode up over a heap of ballast and after a lot of manual digging and the employment of track jacks, the ballast cleaner was edged forward and back on the track. It then carried on with the cleaning. When rerailing crew arrived, it had already finished its operations and was waiting at the end of the line. The foreman in charge of the rerailing crew started to berate me about putting the machine back on the track ourselves, but I countered with, "If you lot had got off your arses and come straight away, you would have been here in time to do the job." He didn't agree but he knew his duty and carried out an inspection of the machine before signing it safe to travel out of the possession.

The problems with the cleaner had caused a delay to the work and I was anxious to monitor progress towards "hand-back", so I stayed on site till late in the evening before jumping into my car to go home. Unfortunately, in the dark I didn't notice a deep ditch and reverse straight into it. I had to shamefacedly ask the relaying supervisor for a gang of men to lift the car out of the ditch, which they gladly did but for weeks after, questions about my driving cropped up frequently.

Possession of the 2-track sections of the ECML and of the Doncaster to Leeds line were of about 12 hours on a Saturday night to Sunday morning and that meant that in order to carry out a full relay and ballast clean and return the track at 125 MPH, the actual lengths of relays were fairly short; whereas on the 4-track sections between Grantham and King's Cross much longer relays could be done because we were allowed to block 2 lines from Saturday afternoon through to Sunday lunchtime when the line would re-open at 50MPH for the Sunday afternoon/evening peak. Later, on Sunday evening, the possession was taken again to prepare the tracks for 140 MPH running on Monday morning. My first handback shift was on one of the latter sections, somewhere around St. Neots. The 50 MPH opening was normally done by the relaying supervisor, but the 140 MPH opening was down to me.

I left home sometime before 3 o'clock on the Monday morning, went to site, did all the checks, signed the line fit to re-open at line speed and then went to the office in Peterborough and did a normal day's work before visiting mum and then going home at about 4pm. I really enjoyed it but there was a lot of pressure, not just to get the checks done correctly but also to re-open the line on time and at full line speed. Incredibly, Peterborough area went for well over 4 years without a single speed restriction on the main lines and I certainly didn't want to be the one to spoil that record. The early Monday morning shift pattern also suited me, so most weekends found me doing that job. Normally it was no problem but if the track quality wasn't perfect then I had a judgement to make but fortunately, I was never on a site where the end result was so bad that I couldn't open at line speed, although it was sometimes a close call and I would drive home wondering if I had done the right thing and half-expecting an urgent phone call at 6am saying the driver of the first train had reported a rough ride!

One such occasion was on the up main north of Huntingdon. I did all my checks and found the actual relay was good but just off the end of the relayed track was some switch and crossing work which had cross-level faults of about 20mm. They must have already been there before the relay started so they were no worse than they had been during the previous week so I called the Permanent Way Maintenance Engineer and told him that I would be opening the line at 140 MPH providing he guaranteed to have a gang to the site that morning to carry out remedial work. I really gave him no choice because if I imposed a speed restriction, he would have had to "carry the can". It is interesting to note that after privatisation, my services as a qualified handback engineer were often in demand but the track conditions that I found then often led me to refuse to open the lines at full speed and on one occasion, having been sent to a site near Scunthorpe to raise a speed restriction over a bridge from 20 MPH to 50 MPH, I actually had to reduce the speed to 10 MPH because of the appalling workmanship.

I am not sure if Alex had some grand plan to convert me into a Permanent Way Engineer, but the next surprise was when he sent me to Doncaster to stand in for the Permanent Way Maintenance Engineer (PWME), John Eland, who was going on holiday for a couple of weeks. The PWME has direct responsibility for the condition of the track and safety of the line on his "patch", which in John Eland's case was a substantial part of the ECML plus most of the Doncaster to Leeds line. I really was out of my comfort zone doing that job. I was still a novice at Permanent Way Work, and I didn't know the supervisors who would be working for me. Furthermore, I knew that John Eland was an old colleague of Alex's, so when I found what I considered to be a problem with some track, I felt that I had to act diplomatically.

I had only been doing the job for a couple of days when there was a report of bad track near Adwick on the Leeds line. I can't remember whether the report originated from a "rough ride" experience by a driver or whether it was detected by the track recording vehicle but either way it needed actioning. I spoke to the P-Way supervisor about it and his reaction was to say that he knew about it but he was waiting for the track to be tamped. Apparently, John Eland also knew about it but was content to leave it alone until a tamper was available. I decided that I ought to check out the track myself, so I visited with the local "ganger" and I was not happy with what I found. The line speed was just less than 100 MPH (if memory serves me correctly) but the track was in poor condition with several significant "twist" faults. The ganger tended to agree with me and said that he had wanted something done about it for some weeks. I had three choices open to me, but I immediately rejected the "do nothing" option which may have been OK for experienced engineers and supervisors who had "seen it all before" but I couldn't have slept at night if I didn't get it sorted. The second option was to impose a speed restriction. Bearing in mind what I have said previously about it being a "point of principle" not to have TSRs on the main line, that option would have meant the shit hitting the fan with a considerable amount of it flung in my direction and possibly also at John Eland for letting get in such a state. The third option and by far the most attractive was to get the track faults sorted, but how?

With no tamper immediately available, I investigated restoring the track by "jacking and packing" using as many men as I could find, equipped with 6 sets of "Kango" electric packing hammers. I arranged for the work to start the next day and the track was soon looking much better. My actions had caused ripples, however, and I soon got a call from Alex who wanted to know what I was doing. I truthfully said that I thought the work was urgently required and he didn't disagree although he probably thought it was an over-reaction. When I next bumped into John Eland after his return to work, he said to me, "I hear you have been doing some work on my patch," and when I said yes, I thought it was necessary, he just "harrumphed" and walked away. However, he didn't actually say it wasn't necessary, and we remained in a good working relationship. My personal satisfaction was that I had got the repairs done that I thought were necessary without dropping anyone in the proverbial.

I enjoyed gaining the P-Way knowledge and experience but had no aspirations to move from "Works" to "Way", however, Alex had other ideas. Our P-Way renewals assistant, Dave Pepper was moving on and Alex asked me to apply for the job in such a way that made it clear I should not refuse and in due course, following an interview, I was appointed to the post. It was a good promotion for me to a position that I had never aspired to. By this time, we were essentially a part of the "Inter-City" business, and I was responsible for the entire track relaying on what we considered to be the premier line in the country. Again, I was fortunate to inherit an excellent staff, both in the office and out on site, so really all I had to do was ensure that everyone was pulling in the same direction and leave them to it. The relaying supervisor at Peterborough was a colleague of mine from the old days, Ray Cousins, and the London end supervisor was named Wyant and was father of Brian Wyant, one of my work's supervisors. Ray's son, Chris Cousins, was also working in the system and is still a senior relaying supervisor for Network Rail.

The week mainly consisted of a Monday morning meeting to review progress on all the previous weekend work then planning the work for subsequent weekends plus, of course, site visits on all weekends. There were also InterCity meetings held in York where managers and staff from all parts of the business got to meet and exchange ideas. I had never known such close co-operation between departments, and I loved it. I think we all thought that we were now working for an organisation that was growing and well regarded.

There was one aspect of the operation of our renewals work that was unique to Peterborough and without doubt, it gave us the edge over any other area. British Rail introduced the "Band 3" system of radio communications throughout the Network in the 1980s, principally to facilitate communication between train drivers and signallers but by the time I moved to Peterborough it was being used widely by the engineering departments for communication with site staff. All supervisors and technical staff carried "Band 3" radios on site and while I was Renewals Assistant, we were issued with a new generation of radios known as "Brunels". The railway's band 3 system operated through line-side masts and coverage on the ECML was pretty good. Each radio had a unique number and could be contacted from other radios or railway land line telephones. It was, however, a one-way system so it was necessary to adopt "over and out" protocols which not everyone found easy. All this was available throughout the system but where Peterborough area went further was to obtain a number of mobile masts that could be erected at relaying sites to provide a local network. Staff using their own "Brunels" were required to re-tune to a new channel when they arrived on site, and this allowed them access to the open site network and be in direct communication with all site staff. It was an invaluable tool which was essential to the efficiency of the operation. The point was that as well as being used for messages, the whole operation of the site was made generally available. For instance, a typical situation might arise where someone would ask where a particular wagon was located, and the reply would come back from someone else that they had just seen it and it was the fifth wagon from the locomotive on ballast train 8M67. All locomotives and on-track machines had cab radios and the drivers were instructed to tune to the site frequency as they entered the possession then all movements of ballast trains and machines was controlled over the radio so, not only did the drivers and supervisors have an excellent tool for controlling train movements, but all site staff knew when ballast train movements were being made. There was also a discrete channel that could be used for direct communication between driver and shunter when safety-critical train movements were made. Overall, I estimate that site efficiency increased by at least 30% as a result of our radio system and if, for any reason, the system failed, then an over-run became extremely likely because we relied on the system to achieve high levels of efficiency and greater overall lengths of track relays.

The system had its detractors, mainly from members of the OHLE who preferred to use mobile phones. I personally believed that was so that they could avoid being at our beck and call. It was very frustrating, though, and one night, it caused me considerable embarrassment. Chris Green, who was still head of Network Southeast at the time had been told the success of our system and wished to see it for himself. So, on a particularly foul night of high winds and pouring rain, I found myself alongside Chris and Alex Turner on the track near Biggleswade, trying to demonstrate how good the system was. I gave Chris a brand-new "Brunel" and he was able to listen to the early site preparations and the possession being taken quickly, but then nothing because for well over an hour, we could not contact the OHLE staff who were isolating the live wires for us. They were not on the "Band 3" system and not answering the mobile phone numbers we had for them. Even an irate call from Chris Green to York Control failed to get anything moving and, in the end, he went home wet and disgruntled. I hope that he took away the right message, which was that for the system to work properly, everyone had to buy into it.

In complete contrast, our own staff and the operating staff really took to the system and often demonstrated how it could really improve productivity and safety. One method of working did push the boundaries of the rule book but demonstrated the effectiveness of the system. Relaying on the 2-track section south of Peterborough was being carried out on the up line with single-line working (SLW) in place over the down line. One of our favourite operations supervisors, Mick "Crackers" Jacobs, was acting as Pilotman which meant he gave authority for trains to travel over the single line and frequently would travel on the loco of the train in question. Since our machines were obstructing the down line as the work proceeded, normally, we would be advised to stop work and lift protection by the signaller and when that was done, the signaller would be told and he, in turn, would tell the pilotman and clear the signals for the train. This would all take at least 20 minutes, each time we had to pass a train.

Mick devised a system which led to minimal delays. He would call out over the radio saying that he was leaving Peterborough with an up train and all tracks were to be cleared immediately and the protection lifted. All the site staff would hear the message on their radios and immediately move clear and the hand signallers standing at the protection would remove the detonator and "STOP" board as soon as they could see (or were told, again by radio) that the line had been cleared. Site wardens were already in place along the length of the relay, and they would warn anyone who might have been tempted to stray onto the live line. Mick would advise by radio when the train was approaching the site and give permission for work to re-start as soon as he had passed. Anyone who has been involved in SLW and knows how long and frustrating delays can be couldn't help to be impressed and the point was that everyone felt safe because they could hear every part of the operation over their own radios. There was, of course, a "fail-safe" because the protection remained in place until the line was clear and if the train arrived before that, it would be stopped by the explosion of three detonators as it passed over them.

We were certainly in the forefront of using the "Band 3" system for engineering works and the radio engineers at the British Railways Board recognised this and would try out new "goodies" on us from time to time. One of these was to provide a facility (located in King's Cross railway telephone exchange), which would enable a limited number of our own house phones to be connected to the system and I was able to contact all our site staff from the comfort of home although it was more often used by the staff themselves to drag me out of bed and off to site myself. It was strange to have to say "over" and use radio jargon over a normal telephone, but my wife got quite adept at it and we had some good conversations because it wasn't possible for her to interrupt me!

As time went on and we made more and more use of radio, the BRB engineers were trying to get other parts of the system interested and they asked me to arrange for a demonstration to be given to Western Region staff at Swindon, which I duly did, supported by a Ray Cousins who arranged for an aerial to be set up in some sidings in Swindon. The demonstration went well but the Great Western distrusts anything they haven't thought of themselves and continued to resent the rest of the railway for not adopting "Broad Gauge", so they were less than enthusiastic. It is also difficult to explain the efficiencies that could be achieved to people who think they are perfect already. Once privatisation came along, systems that required good co-operation between different branches of the railway became almost impossible and the new relaying companies didn't want to be involved with "Band 3", preferring to use ordinary mobile phones so all the progress that had been achieved was tossed away along with a lot of other things which were good about British Rail.

I carried on as renewals assistant until 1993 when our Assistant Area Civil Engineer (Works), Brian Saunders moved on and Alex asked me to take temporary charge of the whole of the works department. Of course, I am biased, but I do believe that the Peterborough renewals section was one of the best in the country and one of the last jobs that I was involved with was the trial introduction of the new "FASTCLIP" fastenings into the main line under the direction of Dr David Cope from Board Headquarters. Needless to say, I took the opportunity to impress on him the use of Band 3 site radio and regular 140 MPH handbacks. Thinking back, though, the reason we were so good had little to do with me and everything to do with the excellent renewals staff at all levels, including the relayers themselves. My main contribution was to recognise this at an early stage and then to give them every support and encouragement and empower them to make their own decisions. My own lack of practical skills as a relayer came home to me some years later following privatisation and will be recounted in a subsequent chapter.

Ballast cleaning on the ECML – Photo Brian Maddison

Chapter 16

New Barnet station fire

I had been back at Peterborough for a couple of months when I found myself dealing with the aftermath of a major fire. I was duty manager "on-call" for the week and late one Friday evening in July 1989, I got a call from control to tell me that New Barnet Station was on fire and that the line was blocked. New Barnet Station is a suburban station on the East Coast Main Line about 9 miles north of King's Cross. It comprises up and down island platforms, each with one face for the slow line and one for the fast line. Access is via staircases from the footbridge located towards the north end of the station. The bridge was wrought iron with timber decking and at the time of the fire, it also carried the booking office and associated buildings. The staircases were wooden. The fire had wiped out the booking office, all the timber bridge decking, and the staircases leading to the platforms.

It took around 2 hours for me to drive to New Barnet and when I arrived, the remnants of the fire were being damped down by the London Fire Brigade, but the lines remained blocked with all traffic diverted via the Hertford loop line. On site was Dick Worby, the assistant Works Supervisor from Hitchin depot who greeted me and put me in the picture. It was suspected that the fire had been started by arsonists; probably local youths and there was little we could do beyond gathering information and summoning resources while we waited for fire investigators to complete their work. This time was well spent, however, and a gang of artisans from Hitchin depot were roused from their beds and brought to site to await the chance to start work. These staff were not "on-call", we relied on goodwill to get volunteers but loyalty and the possibility of some overtime to supplement the rather low wages meant that we rarely had a problem raising a team in an emergency. We also got a couple of lorries onto site and started to make arrangement for a train of empty wagons into which the old, burnt timbers could be loaded.

At about 3 or 4 o'clock on Saturday morning we got a start in stripping out all the old, burnt timber, a filthy job, and while the lads were getting on with that, I started to formulate my plans for restoring traffic and re-opening the station. The slow lines were able to be re-opened as soon as any loose and burnt timbers above them were removed and once this was done, the pressure to re-open the main lines receded and we were able to carry on the demolition of the remains of the booking office. The teams on site could be relied on to remove all the old materials quickly but I needed to have a plan for re-opening the station which clearly involved re-decking the footbridge and re-building the staircases. No attempt would be made to re-build the booking office on the bridge. Later a temporary booking office was installed on the station approach. About this time, I was accosted by an irate Bernard Dawson, who was the Works Supervisor at Hitchin (on whose patch the station was). Bernard said in a very loud voice, "Why didn't you call me, Brian?" and would not be placated by my protestations that he was not on call and that Dick Worby was managing perfectly well. Think about this, dear reader, here we have a railway supervisor complaining bitterly that he had NOT been called out in the middle of the night to attend an emergency on the railway. I like to think that this epitomises the loyalty and spirit of the railway family at the time.

Anyway, once the little altercation was over, I was pleased to have Bernard by my side. He and I got on pretty well, even though he did not suffer fools gladly and needed to be convinced that I was not a fool during the first few of our meetings. We discussed how we could get hold of sufficient timber to re-deck the bridge and to erect handrails. He had some stock at Hitchin, and we could also get timber from Nobby Clark's depot at Palace Gates and from the carpenter's shops at Peterborough. I was pleased to agree all this with Bernard and leave him to look after the replacement of the decking while I started to deal with the problem of the staircases. Just then, my deliberations were interrupted by the arrival of a Senior Operations Manager from King's Cross, I knew him slightly and I also knew that he had a good reputation for getting things done. We agreed on arrangements for re-opening the main lines and what protection was need for the safety of staff working above the lines. In those days the default standard was lookout protection which was used but with some "between trains" protection for certain operations and at night. The operations manager then tempted me with the phrase, "I expect I have lost New Barnet station for a couple of weeks." I can only assume that it was part bravado and part the creed that everything is possible that I gained at first from my father and later from my time with the Army Cadets and on the Outward Bound Course that I attended that made me answer what I did. "Oh no," I said, with complete certainty, "you will have the station back for Monday morning peak, shall we say six o'clock." Of course, operations were well pleased with this, and control was immediately informed. All I had to do now was get 2 staircases built and at the time I had only the sketchiest ideas of how to do this. Nevertheless, I was now committed, and that deadline was then set in stone because train plans would straight away be drawn up showing the first commuter trains stopping on Monday at 6am.

It is now necessary to step back a couple of months in time to when I first arrived at Peterborough as Works Construction Assistant. Prior to moving to Peterborough from Nottingham, I was used to having all engineering small plant being provided "in-house". Furthermore, when scaffolding was required, I had to issue tenders to 3 different companies who would bid for the work. At Peterborough, things were different. Most plant for the works department came from a single plant hire company and it also transpired that most scaffolds were procured from a scaffolding company located in East London. I didn't want to be the new boy wading in to change things, but I knew to continue with these practices could have got me into serious trouble for not instituting competitive tendering. I talked to the bridge assistant Eric Jeyes who saw my problem but also pointed out that the service that the railway got from Prendy scaffolding and Wilsons Plant was excellent and that they had both been supplying the railway for years, so he suggested that I didn't do anything hasty. By the time of the New Barnet fire, I had made some changes to procedures, but we were still using both companies extensively. I was very grateful that I heeded Eric's advice not to change things too quickly without looking at the consequences.

Before 7 o'clock on Saturday morning and long before I could have raised any other plant hire company, Harry Wilson from Wilson's Plant arrived on site to tell me that he had heard about the fire on the radio and that he had opened his depot to supply us with whatever plant we needed. That was amazing service and I found out later, it was not the first time that Harry Wilson had helped the railway in times of emergency. With plenty of manpower, plant and materials available, work was proceeding well but I still had not resolved the problem of the staircases. There was nothing useable left of the original stairs so it seemed to me that the only option was to get a specialist scaffolding contractor to build 2 new staircases which would extend from the north side of the bridge to the platforms, but could I get one to turn out at the weekend? My first attempts to contact local branches of major scaffolding companies proved futile so I rang Eric Jeyes at home for advice. He told me to leave it to him for a while and he would see what he could do. Within a couple of hours, the boss of Prendy Scaffolding, Mickey Prendergast, was on site and sorting things out for me.

Lorry loads of scaffold tubing and boards arrived and shortly after and the staircases were being built.

About 4pm on Saturday a passenger train drew up on the down main platform and I was about to run down to warn the guard not to let any passengers off when one door opened and several hampers of food were unloaded, courtesy of the operations manager at King's Cross who arranged for the station buffet to be raided and the food sent to our staff. We were very grateful for the gesture which rather summed up the way that the railway of the time worked together at times of need. By that time the footbridge decking was well advanced, the scaffold staircases were being built and temporary fencing and handrails were being erected. Painters were even on site to colour handrails white and highlight any obstructions. Temporary lighting was being installed by the Plant and Machinery Department and most of the old timbers had been loaded away to tip.

Shortly after that, I left for home, confident that work was proceeding. The station was open for the Monday morning peak, and I would have loved to have seen the faces of the commuters who travelled home on Friday night through the old station but who arrived to a completely changed station on the Monday morning. Furthermore, the scaffold staircases and other works that we carried out remained in use for over 10 years before they were replaced by new, permanent structures as a part of a station upgrade.

New Barnet Station before the fire. The large building on the bridge was completely burnt down together with bridge decking and staircases.
Copyright Martin Addison

New Barnet Station today, showing the area where the booking office once stood

Chapter 17

Chartered Engineer

My education and training before I joined the railway could best be described as "could do better" so I was quite relieved in 1964 when I was told that the Building Construction Course that I was following at Peterborough Tech College would be of no use to me and that I would be re-enrolled for an Ordinary National Certificate Course in Civil Engineering. This enabled me to draw a line under two years of study during which I abjectly failed to understand what I was being taught and spent much of my time either on my motorbike or in the pub. This was partly due to my own shortcomings but also because I was the only member of the class who was not putting into practice what I was learning at college when I was back at work. My fellow pupils were mainly trainee architects and budding building site managers whereas my only connection with building construction was ordering the materials and checking invoices. I got so far behind the rest that I skipped classes which of course made things worse.

What a difference when I started the Civil Engineering course! By then I had been on the railway for 4 months and I was already carrying out the work that I was to be taught at college. In that respect I was ahead of the rest of the class instead of being behind. This made things much easier in all subjects except maths, which continued to be a bit of a mystery to me. Tim Green, my immediate superior was a Chartered Engineer and he had gained his Membership of the Institution of Civil Engineers (ICE) via the same route that I was now embarking on, and I had high hopes that by my mid-20s, I might achieve that goal. Because of my previous time at college, the ONC course was reduced from 3 years to 2 and I progressed well in most subjects. When it came to taking the actual ONC exams, I found that the structural design examination clashed with my Outward Bound Course in Scotland but the college arranged for me to take the exam in Burghead under the supervision of a member of the staff there. Since the exam involved practical drawing work, the school also found me a drawing board and Tee-square.

Unfortunately, although I did well in most subjects, maths again let me down and I was referred in that subject, meaning that I had to repeat a full year doing maths only. Although this meant that I lost another year in my target of achieving a Higher National Certificate, by concentrating on maths only, I eventually passed and got my ONC. During the "maths" year I had to go to college on one afternoon and one evening each week which was somewhat inconvenient to my employers but was great for me. By then I was working in King's Cross so soon after noon on a Thursday every week, I left work and joined an express train back to Peterborough. Most days I went straight to the restaurant car where for fifteen shillings, I could get a 3-course lunch which I finished just before we pulled into my destination! My favourite was Gammon and Pineapple – very '60s food! It was about this time that the ICE dropped their bombshell and announced that in future, the minimum academic qualification for a Chartered Engineer would be a degree which, of course, was a great disappointment to me. The alternative offered was to become a Technician Engineer (later known as an Incorporated Engineer) and I determined to achieve this even though I considered it to be a rather "second class" qualification.

Before starting the first year of my HNC course, I moved to Leicester and set about finding a course there for myself. Unfortunately, my search in Leicester proved less than fruitful since being a town of light engineering factories; most of the courses were in production engineering and other related subjects. I eventually had to enrol in Kedleston Road College in Derby which involved a return journey by train and bus (or a long walk) of over an hour in each direction. I have a dislike for Derby which persists to this day and probably has its roots in those long winter journeys but then was exacerbated by football rivalries when I became a Leicester City fan (Derby were the more successful team at the time). The one thing that I did have available was time because I was living alone in a bed-sit in a city that I didn't know, so instead of going to the pub every night with my mates as I would have done in Peterborough, I had nothing better to do than settle down with my Leonard Cohen records and study.

This paid off immensely and not only did I progress well in all the engineering subjects, but I suddenly understood the maths! It was like a revelation to me, calculus which had previously been a real mystery became crystal clear, as did all the other parts of the maths course and although the time spent studying reduced as I made more friends in the neighbourhood and got into relationships with the opposite sex, it still seemed like all the barriers had been broken. I went on to complete the HNC course with distinction and added 4 more advanced subjects to the basic HNC. In my final year I was awarded the British Constructional Steelwork prize for my work on structural design and at the end of that year, I was approached by the college to see if I would enrol on a full-time degree course but by then I was married with a child on the way and I also felt that after 20 years of study, I needed a break. I went on to achieve Incorporated Engineer status and then put aside all further thoughts of academic advancement.

Many years later, probably in the mid-1980s, I was made aware of a scheme being launched by the ICE that was intended to allow experienced engineers to become chartered who had missed out during the early part of their career. I looked into it and found that although there was a lot of work involved, it was possible that I might qualify but I did not act immediately. Only when I moved to Peterborough and had the opportunity to discuss the idea with Alex Turner did I decide to pursue the matter. Alex was immediately supportive and in 1991, while I was still Works Construction Assistant, I made my initial application to the institution. By then I had produced the necessary volumes of supporting information and obtained the sponsorship of 4 Chartered Civil Engineers, including the recently retired Chief Civil Engineer of the Eastern Region (P. B. Davis) and the Chief Civil Engineer of the Scottish Region (R.J. Spoors). It took a further 2 years, but I was finally admitted to the Institution of Civil Engineers as a member on 17 December 1993.

In order to gain membership, I had to produce detailed evidence of the work that I had done over the years, and this amounted to several volumes of text, photos, drawings, and calculations. All that was relatively easy, although time-consuming. I also wanted to include a "flagship" project that would demonstrate my ability to follow through a complex job from start to finish and severe winter flooding in 1992 provided me with exactly the opportunity that I needed. In the 1980s, riverbed scour had become of major concern to railway engineers following some bridge collapses due to the material under the foundations being undermined by "scour" during times of flood and on many bridges, markings had been painted for the guidance of engineering staff during flooding. They were a blunt instrument and tended to be rather conservative since water level did not always correlate with river flow, but they did prove useful.

On Bridge 144 at Huntingdon, we had a yellow line which mandated the introduction of speed restrictions if river levels rose to touch it, and above that a red line which indicated that the railway should be closed immediately and remain closed until the foundations had been inspected by a diver. I have already mentioned Bridge 144 in connection with the Network Southeast paint scheme that we had applied, and you will, perhaps, recall that it carried the East Coast Main Line over the River Ouse. Any closure of the bridge would, therefore, cause huge disruption to passenger and freight services. Furthermore, since the River Ouse was generally a benign and controlled river, the closure would almost certainly prove to be unwarranted and poor old British Rail, who could rarely do anything right in the eyes of the public, would be severely criticised. That winter, the local patrolman reported that the Ouse was rising above normal levels, and it was arranged for regular monitoring of the height relative to the markers. I was not only Works Construction Assistant at the time but also the only diver at Peterborough since I retained my qualifications and membership of the Nottingham team. The water rose to reach the yellow line and speed restrictions were applied and a permanent watchman was appointed but the river continued to rise until it was almost touching the RED.

I accompanied Alex Turner to site, and we watched the river flowing fast under the bridge; if the RED marker was reached, Alex would be presented with a difficult choice – did he stop the line and cause major disruption or continue to monitor the levels in the hope that they would soon fall, which was the prediction from the Environment Agency. We consulted, and my view was that the river flows, although fast, was not extreme and that undermining was unlikely to be taking place. Furthermore, I volunteered to "take a look" underwater to see if I could get close enough to the piers to check the foundations. The course of action agreed was for me to get some urgent support and equipment from the Nottingham Divers to carry out the examination as soon as possible (within a few hours) and in the meantime the condition of the bridge and height of the river would be monitored from the bank. The Nottingham team had previously developed a technique for examining bridges following flooding but while the rivers were still flowing fast. Once they arrived, I got all my kit on, including SCUBA and walked about 50m upstream from the bridge, carrying a metal "prodder" which was about 1.5m long. Armed with this I launched into the flow of the river, aiming for the nose of the nearest pier.

I was bowled along at the mercy of the current but was able to grasp the cutwater and force myself down to the bed of the river in the slight slack that was created. I was then able to prod the bed at the point where it was next to the pier and follow this line along the length of the pier before being carried off downstream in the current. It sounds a dangerous operation and my Peterborough colleagues watching from the bank thought that I was lost when I disappeared from sight but in fact, it was a technique that we had practised many times and was quite safe as long as I followed the golden rule and didn't try to fight the current. Instead, once I reached the other end of the pier, I simply let the current take me while I gradually worked my way to the bank. Once there, I hauled myself out of the water and trudged back to the bridge. I repeated this operation on all of the piers and reported back to Alex that there was no undermining, so he was able to remove the speed restriction as soon as water levels stabilised and we had avoided shutting the bridge. I was very satisfied with the outcome but it was clear that such an important bridge needed better flood protection so that the situation would not be repeated. Alex raised the matter with the InterCity team at York who readily agreed that money should be spent, and I was appointed to be the engineer for the job because of my knowledge of underwater matters and I was able to make it my major project for my M.I.C.E. application portfolio.

My idea was to start with a detailed inspection of the piers, once the river had dropped to normal levels and then to carry out whatever site investigations were required before designing and installing appropriate scour protection works. For it to be a valid project to submit to the institution, I would need to provide proof of the efficacy of the scheme at all stages of the works but at the same time, the works were required to be completed before the winter of 1993. Fortunately, in 1991, prior to the flooding, the piers had been drilled to determine the foundation depth, soil sampling had taken place to determine the bed material and the Hydraulic Research Station at Wallingford had carried out a scour risk assessment. All this, together with a detailed underwater examination gave me the information to proceed with design works. Initial considerations led to me preparing schemes for providing steel sheet piling protection and for providing over-bed protection in the form of large stones or stone-filled gabions (known as Reno mattresses). In the end, the Reno mattresses were chosen, both on the grounds of lower cost and ease of installation, but all schemes were fully designed and costed. This formed a major part of my M.I.C.E application and although when the documents were returned to me from the institution, they were covered in pencilled corrections and comments, nevertheless my application was considered sufficient for me to proceed to the final stage which was for me to be interviewed by a panel of senior fellows of the society at the institution's headquarters in London. By the time the works were implemented on site, I had moved on to being the Renewals Assistant, but I retained an involvement in the works which were installed by staff from the works depot at Peterborough with assistance from contract divers.

Eventually, I got the summons to Great George Street to be interviewed by a panel of three engineers: one was to be from the rail industry, one from the marine side of the business, and one other. On the day in question, it snowed heavily in the West Country and two of the engineers failed to arrive. One had not even been able to leave home, and the other was on a train that was heavily delayed, and he reported that as soon as possible he would be heading home. However, the interview was not cancelled, and I began to believe what I had already been told by some of my colleagues, that if I got as far as the interview, I was 90% there. The sole remaining interviewer was the marine engineer but his background was in ships and harbours, so he was not experienced in river works. However, he had read my report and pronounced it satisfactory, and we then had a very pleasant discussion about marine corrosion and its effect on ship design, to which I contributed little but learned much. Eventually, I was awarded my membership of the institution and the certificate hangs on the wall of my home "office". It was the pinnacle of my academic achievement, and I was justifiably proud of my success. It was also essential to my permanent appointment as Works assistant which I describe in the next chapter. So, 29 years after I embarked on my career as a Civil Engineer, I finally made it to be fully qualified and have the letter C.Eng. and M.I.C.E. after my name. I like to think that my dad would have been proud.

Bridge 144 ECML – Brick piers in the River Ouse

Chapter 18

Works Assistant for Inter-City

For the whole of my time with British Rail, the engineer in charge of a particular district, division or area was provided with a P-Way Assistant and a Works Assistant who would head up their respective sides of the organisation with the support of a Chief Clerk. Titles would vary but by 1992, InterCity was operating as an independent business and so the Works Assistant at Peterborough held the title of Assistant InterCity Area Civil Engineer (Works), Peterborough. At least I think I have the right order of the words but from here on I will refer to the post of "Works Assistant". Since both assistants would, from time to time, have to deputise for the engineer, it was deemed essential that the occupant of the post should be a chartered engineer. Alex asked me to occupy the post of AACE (Works) as a temporary measure from early in 1993, but I could not be formally appointed until I became chartered later that year. I was "over the moon" as footballers say when they score a goal to be appointed to the post which more than exceeded the level I expected to achieve at that stage in my career, however, my joy was somewhat tempered by the announcement from John Major that the railways would be gradually privatised from 1995 onwards and that InterCity East Coast, being one of the most profitable parts of the business, would undoubtedly be one of the first to go.

In the event, privatisation was a more brutal process than I could possibly have imagined but in 1993, we were looking forward to the event with a combination of trepidation and excitement. The organisation that I took charge of was, to some extent, my own works department that I had moulded together when I was Works Construction Assistant and all my supervisors were still in place. In addition, the heads of building and bridges section reported to me. InterCity was largely self-contained in all its resources, and this extended to its own fleet of locomotives, wagons, and on-track plant so I had access to a very large range of resources. But what I didn't have was a sufficient budget, either to pay my staff or to maintain the structures as well as I would have liked. Looking back know, it is incredible what was achieved by works departments throughout British Rail. We kept the railways safe and running despite having very limited budgets. Oh, how I wish that I had the sort of money that has been lavished in the railway system since privatisation, almost all of which has come from the public purse. King's Cross station is a typical example; it has been magnificently restored at a cost in excess of £500 million pounds and the results are fantastic but that sum of money would have covered the entire budget that I had for ALL structures on my patch for over 100 years and my budget for King's Cross Station for many thousands of years. So, when people compare the King's Cross of my day which could be somewhat dingy but definitely functional with the King's Cross today, just remember how much it cost.

My task was threefold. Firstly, to keep the structures safe using the limited maintenance budget that I had; secondly, to find enough work beyond day-to-day maintenance to keep all my staff gainfully employed. and thirdly, to prepare my organisation for privatisation. Keeping the structures safe was a question of ensuring the reports received from bridge and building examiners were acted upon correctly and promptly. The examiner's reports were first seen and commented on by Works Supervisors and they made their recommendations and then by the bridge and building assistants before I received them, but the work recommended was inevitably more than the budget would stand so I spent many late nights assessing the reports and the comments before adding my own recommendations and priorities. The final decisions were then taken by Alex, whenever possible during a formal tour of the structures by the Engineer's Inspection Saloon or by road. Once work was authorised, I had to make sure that it was carried out properly and safely. All that was pretty mundane but absolutely essential at a time when many senior railway engineers were being deflected from their main duties because of the forthcoming privatisation. I felt keenly that I and my colleagues on other areas a similar level were the ones who had to look after the day-to-day maintenance and working of the railway system, making sure that it continued to be safe and fit for purpose. I was helped by the excellent team of supervisors, bridge examiners and managers but in what would later become very difficult circumstances immediately prior to privatisation, it was an onerous task.

Previously, I mentioned trips on the "saloon" to review proposals for work and in my role as Works Assistant, I accompanied Alex on these trips. Very few could be organised on the ECML because there was simply no space in the timetable for it to run. Most of those visits, therefore, had to be done by road, but just occasionally we got the saloon out and so, one day, we all found ourselves at Doncaster Station waiting for the it to arrive and take us down the Leeds line. It eventually came in, a little late but with a very cheery driver who set off with aplomb. The duration at each site was severely limited because there was always a train "up our arse" as the saying goes but we had a good day and completed the Leeds line then headed off for Hambleton Junction on the ECML via some interesting freight lines. Once there, we stood for some time, awaiting access to the ECML and eventually the driver told us if we waited much longer, his day would be up, so he offered us a trip home via Selby. We had little choice but to accept, although it meant leaving several sites for another day and we had a good (and fast) trip back to Doncaster via Selby. The days when alcohol was served on the saloon were long gone and that is probably why I have no especially interesting anecdotes to relate about the trip.

I like to think that my chief attribute to part one of my task list, was to insist on all the procedures being followed that I knew would keep structures safe; at a time when many people were being drawn into the privatisation procedure, I knew that any defects in our structures were being attended to. Unlike other areas at the time, I insisted that the bridge and building examinations were not allowed to lapse and that the "farming out" procedure was continued with because the latter ensured that authorised work was monitored until it was completed.

Task 2, keeping everyone gainfully employed, meant finding projects to work on that were not included in the maintenance or renewals budget, so we went looking for new works projects and work on other areas. The number of new projects on our own area was severely limited, although we did find a few, such as constructing a public access to a former parcels subway on Doncaster Station and constructing long lengths of security fencing around sidings in the London area to keep coaches safe from vandalism. We had more success, however, with Network Southeast and a chance meeting with a manager of the London, Tilbury and Southend Line proved very productive. Apparently, they had a sizable budget that had not been spent because Stratford area did not have the resources to spare. We jumped at the chance and carried out some quite large projects that included repairs and renovations to the inspection pits at East Ham train depot and refurbishment work at Barking Station. The latter involved platform and subway repairs, much of which had to be done on nights. We put together 2 gangs, one from Palace Gates Depot with George Lovell in charge and one from Peterborough with Russ Pettit in charge. The Peterborough team had a fair distance to travel but it was a straight run down the A14 and M11. The teams each took day and night shifts on alternate weeks and some intense rivalry developed between the two. This was good for progress at first because each tried to outdo the other but as time went on, it became clear that healthy rivalry was developing into something less useful. I got reports from each camp of work being left unready by the other shift and even some suggestion of sabotage. I went to site one evening on shift change and laid the law down. After that there was little more trouble.

Work proceeded like this until well into 1994 and I was not only keeping up with maintenance work but also proud of some of the other projects that we had carried out. Now it was time to prepare our staff for the privatisation process. All engineering departments were hived off from the respective business sectors (InterCity, Regional Railways, NSE and Freight) and bundled into one unit called British Rail Infrastructure Services (BRIS). Our particular unit was called British Rail Infrastructure Services (East Coast) with a Head Office at York headed up by Dave Dogget and area offices at Peterborough and Newcastle. At first it seemed an exciting prospect and we were encouraged to seek work in the private sector so as to prove our worth prior to privatisation. Alex's opposite number at Newcastle was Colin Wheeler and for some reason, he considered himself the leader and promised that his works staff would be re-roofing hospitals and other public buildings within a few months. This promise, however, proved to be all "Piss and Wind" and as far as I am aware, they never did do any private sector jobs. That didn't stop Colin travelling down to Peterborough to tell us how good Newcastle was at their job and, by inference, that we were second best. I let him carry on, it was not, in any event, my place to contradict him but while he was blowing his own trumpet (something that he was very good at) we were quietly working on a project of our own.

At Peterborough, there is a preserved steam line called the Nene Valley Railway (NVR). It extends from a station in Peterborough, near to the bridge that carries the ECML over the River Nene through to Wansford Station and on to Yarwell in Northamptonshire. The railway is connected to the British Rail network via the Fletton Branch at Fletton Junction. One of the people associated with the railway worked for Civil Engineering Consultancy and was also a member of the Permanent Way Institution (of which Alex was the chairman of the local branch). He told us that he was involved in the design of a new platform at Orton Mere Station for NVR and he came into the office for some information on certain details of platform construction. In the discussion, I indicated that we had the resources to build the platform and asked if it would be possible to tender for the work. He was very keen for us to do that and we were allowed to pre-qualify for the tender list.

All this was happening just before our transfer away from InterCity and when we provided our tender, we also made an offer to have one of our best Class 47 Locos named "Nene Valley Railway" and that we would arrange for it to visit the railway for a naming ceremony and for the opening of the new platform. We also offered to deliver all the large materials for the job by rail, thus minimising any damage to the Ferry Meadows Country Park, in which the station stands and, of course, we emphasised our experience and expertise in building station platforms. In the event, our tender was found acceptable to the NVR and work went ahead. We employed sub-contractors to install mini-piles to support the platform and all other work was carried out by Peterborough Depot staff. We completed all the work on time and made a small profit on the job. Our client or at least the consultants acting for them were pleasantly surprised that we did the job for the originally agreed price. Only when I moved into the private sector myself was I to discover that contractors made money not by doing a good job at a good price but by putting in outrageous claims for extra payments and that terms like "unforeseen ground conditions" were banded about and argued over ad nauseum.

Unfortunately, by the time the platform was finished, we had ceased to be InterCity and we had lost our own fleet of locos. We were now BRIS and our locos for engineering works were provided from Trainload Freight at Immingham or Toton. They eventually and very reluctantly agreed to prepare a Class 31 and send it down to Peterborough for the ceremony. When it arrived, spraying with EXMOVER and several trips though the washing plant had not only removed the grime but also much of the paintwork from the sides of the loco so that we could not possibly present it in that condition. Paul Moffatt and the Peterborough painters came to the rescue. He sent down to the local paint merchant (Manders in Cowgate, as it happens) for a supply of suitable paint and his gang repainted the whole of the loco in approximately the right colours. When finished off with white lined wheels by the Nene Valley staff, it looked good and RAIL magazine did comment that it sported and interesting "grey stripe" livery variation. The opening ceremony was carried out by Jim Cornell, head of BRIS and our loco stood alongside steam loco "Hinton Manor". We were proud of our achievements and thought that it might herald a bright future although that was not to be. It did, however, give me the satisfaction of letting Colin Wheeler know what we had been able to achieve.

Throughout 1994 we strove to prepare ourselves for privatisation and at some point, BRIS brought in a character called David Swallow and based him at Derby but with some jurisdiction of us at Peterborough. Exactly what his official position was, I am not sure, but it soon transpired that he was a "hatchet man".

One day Alex and I were summoned at short notice to Derby to see Mr Swallow and we travelled in the same car. Traffic was bad and we were clearly going to be late when we spotted a roadside coffee counter. Alex was already pissed off at being summoned to Derby by "some upstart" and he declared that we might as well be hung for a sheep as a lamb and we enjoyed a break before continuing, very late, to Derby. I can't be sure but I think that was the day that we were told that our works organisation was to be disbanded and all my artisan staff and supervisors would be made redundant. We would also be prevented from any sort of management buy-out because there would be no organisation left to buy. It was pointed out to us that there were plenty of contractors out there in the industry that could do our work better than we could, so we were all to go. Of course, they couldn't do the same with track maintenance or renewals so those organisations would be kept together and sold off in complete units, but we were not given that option.

I was devastated, not particularly for myself but for all the staff who had been promised in good faith that the best way to protect their jobs was to work hard and well and prove themselves fit for the private sector. They had been sold down the river and there was nothing I could do about it. I felt terrible. Shortly after that I put in a request for redundancy and since the whole point of rail privatisation put the railways in the hands of private companies, this was readily approved. For some months, I had been wrestling with my conscience because I had been meeting with 2 other members of the diving team who were keen to set up their own company but could only do so if I joined them. I was very drawn to the idea but while there was a chance of preserving my works organisation and transferring them to the private sector in an orderly way I wanted to stay as Works Assistant. David Swallow's announcement made up my mind for me, but I still felt as if I was letting friends and colleagues down.

It was now approaching the end of 1994, but Mr Swallow had one last bombshell for me. He came up to me in the office and said (almost shouted) to me that he had stopped me getting my redundancy payment because he had heard that I was intending to form my own company and "he wouldn't provide me with the capital for my venture". It was a devastating blow because the £10,000.00 of redundancy money was earmarked for setting up our new company and his attitude put our whole project in jeopardy. I agreed with my two partners that we had come too far and that if he made good his threat, I would just leave anyway and we would make do but it would have made our task much harder. I was determined to fight the decision and fortunately, I had the promise of redundancy in writing. I was also fortunate that my reporting line to the top of BRIS went through Alex Turner to Dave Dogget at York. I was able to talk to Dave about the problem and he contacted Jim Cornell (head of BRIS). I am grateful to both of them that their response was, "We have made Brian and offer and we will keep our word," and David Swallow was over-ruled. I kept well clear of him then until I left on 31 March 1995, fearful of him getting his own back because he appeared to be a vindictive man. On the day I left, I was presented with a certificate acknowledging my 30 years of service for British Rail although, disappointingly, it is headed British Rail Infrastructure Services, not just plain old BR.

An inspection saloon, similar to the one used at Peterborough

RAIL

No. 224   APRIL 13 – 26, 1994   £1.70

Nene Valley Railway name for 31558

Above: With daffodils in full bloom, InterCity ECML infrastructure locomotive Class 31/5 No. 31558 makes a fine sight in the Nene Valley Railway's Wansford Yard on March 23. Repainting of the 'Brush 2' - with an interesting 'grey stripe' livery variation - had just been completed in preparation for its naming *Nene Valley Railway* by BR Infrastructure Services MD, Jim Cornell, at Orton Mere later that afternoon. Following the ceremony, the loco hauled a VIP special over the line to Yarwell and back. The Type 2 will be a regular sight passing through *RAIL*'s home city, but now under the jurisdiction of Trainload Freight SouthEast in a Toton allocated, Peterborough outbased pool. It will not be short of celebrity sisters, No. 31116 *RAIL Celebrity* having also made a move to Peterborough in the same week.

Right: The naming also heralded the official opening and dedication of the newly constructed second platform at Orton Mere station on the seven and half mile long private railway. Financed by a bequest from the late Geoff Humpfrey, a long serving NVR volunteer, the platform was constructed ahead of schedule by Peterborough InterCity Civil Engineers. BR's Class 31 No. 31558 *Nene Valley Railway* stands alongside preserved GWR 4-6-0 No. 7819 *Hinton Manor* prior to the banner breaking.
Photos. Philip Sutton.

Article about Orton Mere Station as it appeared in
Rail Magazine courtesy of Bauer Media

201

Chapter 19

Diving – The Early Days

Up until now, most of this book has been concerned with my main employment on the railway as an engineer, but there was also a second job that I have only briefly mentioned so far. That was my role as an inspection diver in the rail industry, a job which spanned much of my working life, both with BR and in the private sector. Even the idea that I was a diver for British Rail was considered hilarious by some people but there is no doubt that as well as providing me with another interesting aspect to my employment and some useful additional income, it also provided a fair few lighter moments. The beginning of the story starts back in about 1962 in Derby, before I even joined BR. The railway companies had always employed their own "Standard" divers in the ports and harbours and BR continued to do so. "Standard" divers were the sort that wore the heavy canvas suits and large copper helmets. They were supplied with air from the dockside by a hand pump and were most useful for carrying out repairs to harbour walls and timber piles. However, by 1962 Self Contained Underwater Breathing Apparatus (SCUBA) which had been pioneered by the likes of the famous French marine explorer Jacques Cousteau was commercially available and sports diving using this apparatus was becoming popular. The District Engineer in charge of Derby South, Mr. M.C.B. Johns, was a keen diver and he saw the possibilities of using it for the underwater inspection of structures because of its light weight and portability. He asked for volunteers amongst his own staff and set up a team which included Alan Lodge, Johnny West, Brian Garner and Gordon Issott, joined later by Geoff Barker. Sadly, Mr Johns, who I am told was a brilliant engineer, fell ill and died so I never got to know him. However, the diving team that he started flourished under Alan Lodge and when I had been at Leicester for a few months, Geoff Barker said Jonny West had left (or been drummed out) leaving a vacancy and would I like to give it a go. Of course, I jumped at the chance and, in the winter of 1968, I was sent to the headquarters of Messrs Siebe Gorman Ltd. at Chessington in Surrey.

Siebe Gorman were renowned makers of "Standard" diving equipment and ran courses for commercial divers (both Standard and SCUBA). I was sent there by BR along with a colleague, Soames Jeanne for a one-week course (it takes at least 10 weeks now). Soames and I travelled down by train on the Sunday evening to the digs that had been arranged for us and then went out for a few pints where we discussed how we were looking forward to the course. The next morning, we had a rude awakening when we were met by possibly one of the most obnoxious people I ever came across in my working life. I can't remember his name, but it might have been Danny, so I will call him that. He was an ex-Royal Navy diver whose pet hates appeared to be women, students, and civilians in general (in that order). He had a real chip on his shoulder and unfortunately for them, 2 of the 5 people on the course were trainee Marine Biologists from Liverpool University, one of whom was female. Strangely enough, the other diving instructor, Bill, also ex-Royal Navy was one of the nicest people you could wish to meet but he was mainly concerned with the training of "Standard" divers.

The five of us: myself, Soames, the 2 marine biologists and a young man who was paying for his own course so that he could get employment as a commercial diver, assembled in the classroom in front of Danny and after about half an hour of ranting and sarcasm he advised us that our first morning would be spent in the compression chamber to see (in his words) "if we could stand it". He also told us that since he had only a week to train us, we all had to complete the tasks that we were given each day, or we would immediately be thrown off the course. Compression chambers are large metal cylinders that are used to put divers under the same pressure that their body will encounter underwater, and they are often used for therapeutic de-compression to treat cases of the "bends". One feature of them is that entry is through a very small hatch and unfortunately that was when Soames first discovered that he suffered from claustrophobia and couldn't face going through the narrow opening. I was already in the chamber when this happened, but I later found out that he got no sympathy from Danny, was called useless and thrown off the course. I didn't see him again till I returned to work a week later. That left just four of us and we all got through the compression chamber test, though in my case, not without very sore ears caused by the rapid changes in air pressure.

The rest of day one was spent in the classroom so after a night alone in the digs I returned next morning to yet more of Danny's goading although he largely ignored me and saved his venom for the university students. That day was spent diving in the 3m deep tank, which was like a small, round (and very cold) swimming pool where we all got used to using the SCUBA equipment. Towards the end of the day, we took it in turns to find out what it was like to work in dark, murky water by putting on a blacked-out masks. We then had to re-assemble a flange joint from the component parts, nuts and bolts that Danny threw into the tank. I hung back a bit so that I could watch how the others fared and they all struggled to find the parts which was the difficult bit, the actual re-assembly using a large spanner was easy enough. Danny was timing each diver and had set a time limit on the task with threats of being thrown off the course if you took too long.

When my turn came, I realised that I had to carry out regular sweeps of the tank bottom rather than swimming about aimlessly and I achieved that by hooking my foot under the bottom of the entry ladder and pivoting about that while I did sweeps for the parts. That really helped and I was quickest at the task. The two students were slowest, much to Danny's glee although he did not make good his threat of expulsion. On the third day, we were all due to go to a flooded quarry for open water training but the couple from Liverpool had had enough and they didn't turn up. I felt really sorry for them; they would have easily made it through the course with just a little bit of encouragement from Danny and I overheard Bill telling him so in no uncertain terms. I think at that stage, Danny realised that he couldn't afford to lose any more from the course and he toned down his attitude considerably.

That day gave me a taste of the extreme cold that divers regularly suffered from. The water in the quarry was turning to ice around the edges when we arrived and although we wore dry suits which kept our body reasonably warm (if we kept moving) our hands and feet were very cold indeed. We put up with the discomfort, however, and at the end of the week I was beginning to enjoy myself. The highlight of the rest of the week was the free ascent in 12m deep tank. This was an approximation of the test given to submariners at the time and we were first shown a Royal Navy training film. We descended to the bottom of the tank wearing our SCUBA gear and we then had to take it off and carry out a controlled free ascent. The important thing to remember was to breathe out all the time during the ascent because the air taken into our lungs at depth would expand as we rose, and pressure reduced, and it was impressed on us that clamping our mouths shut would cause untold damage to our internal organs. In the end I got my cherished certificate which still hangs on the wall of my home "office" and felt ready to join the team although I was to find out later that I was still a rookie.

Diving inspections were normally carried out in the summer when water was warmer, and river flows slower, so I expected to have some time to wait for my first working dive. However, a special job came in just after Christmas and one day, I found myself on the banks of the River Trent next to the Nottingham Forest football ground (Boo). The disused railway bridge over the Trent (now known as Ladybay Bridge) was being handed over to the local authority to be converted to a road bridge and they wanted an up-to-date examination report on all parts of the structure, including underwater. The River Trent at the time was home to many coal-fired power stations that constantly pumped warm water into the river so it stayed warm all winter. On the day we examined it, it was also flowing slowly so I had a very enjoyable dive which lulled me into a false sense of what was to come.

In March 1969, the inspection season started in earnest with a bridge over the River Derwent in Belper, Derbyshire. The river was flowing fast, swollen by heavy rain from the Derbyshire hills and with no comforting power stations to warm it. Full of confidence, I donned my gear and was the first to jump in. I was immediately carried away by the flow and lost one of my fins (as we divers call flippers). By the time that I got control of myself, I was 100 metres downstream of the bridge and cold water was already seeping into my diving suit through the waist joint which I had neglected to seal properly in my rush to impress! I got no sympathy from the rest of the team who were all having a good laugh and I was put straight back into the river to do a proper inspection. I learned a lot that day but probably not enough because there were plenty more interesting encounters with water hazards throughout my career. The equipment that we had at the time comprised a 2-piece rubber and canvass diving suit which joined at the waist by rolling the skirt on the top part with the top of the bottom part and then securing it with a rubber cummerbund. This joint frequently leaked but the worst points were the neck and cuffs which were tight rubber and either leaked or were so tight that wrists in particular became sore and swollen. It was also difficult to get the air out of these suits as they had no "dump" valve so it was necessary to crouch down in the water and open the wrist seals to release the air, otherwise it would either be impossible to descend, or air would be trapped in the legs of the suit which could turn the diver upside down into a very dangerous position. The breathing apparatus consisted of a cylinder of compressed air fitted with a "Mistral" valve and a two-pipe air delivery to a mouthpiece. All quite modern for the time but a world away from the comfortable suits and single-hose regulators of today.

From 1969 to 1971, I was living in a flat in Leicester and the diving inspections for those two years fell into a comfortable pattern of easy work and enjoyable company. Being based in Leicester, I was not exposed to the "office politics" which was to plague the BR diving team throughout its existence but I was aware that some years before, the team had travelled to Wales and other parts of the country to carry out inspections on some of the railway's major structures but that we were now restricted to the Nottingham Division and that the work had to be done in "overtime". Later I learned that "spanner man" had stirred up the other managers who were refusing to release their staff from their normal duties to go diving on other areas.

Consequently, the annual programme was not onerous and could be carried out on a series of Sundays throughout the summer months. The equipment was kept in a store cupboard at Derby Friargate workshops where Gordon Issott was based, and we would meet there each Sunday morning before starting out on the day's work. Geoff, Gordon, and Alan Lodge were based in Derby but Brian Garner lived in Syston (near Leicester). Alan Lodge, being a manager, could get hold of a van for the Derby crew but I had no access to a vehicle in Leicester so Brian Garner would pick me up from my flat in his works lorry. Brian was a roofer by trade and had also trained and worked as a bridge examiner for some years but was now working as a lorry driver. I have never found out why he had taken the step of down-grading from Bridge Examiner to lorry driver, but I surmise that it may have been due to some misdemeanour or because access to a lorry may have suited his out-of-work activities, which were legendary. Brian was the best diver that I ever worked with and could be relied on absolutely underwater, but he was not good on discipline or timekeeping.

One Sunday morning on the weekend that "clocks go forward" Brian arrived at my flat over 2 hours late and explained that he had put his clock back the previous night by mistake. Only much later did I work out that if he had done that, he would have been 2 hours early, not late! I have always been a stickler for timekeeping, so his lateness really annoyed me. Furthermore, he would make matters worse by wanting to go home early on the principle that he "didn't want to be late twice in one day"! Once we met up at Derby, we would have a cup of tea and load up the van but either Brian or Gordon would always have some additional jobs to do and would slope off into the workshops. These workshops had been built by the Great Northern Railway to service their routes in the area which included the line between Nottingham and Derby Friargate. Friargate Station had long closed, but the old "shops" remained in use. They included a fabulous carpenter's shop which had a massive circular saw used for cutting large timbers for bridges and track works. The machinery was well-maintained but antiquated and many of the machines were driven by an overhead shaft and belt system from a steam engine. The boiler for the engine was heated by a fire of wood shavings swept up from the floor of the shop and tended by a lady worker.

The workshop supervisor, Reg Pulfrey, was an excellent man but he had lost a leg in the Second World War when the tank that he was driving received a direct hit. Nevertheless, he was pretty nimble on his feet and not much went on that he didn't know about. However, he didn't work Sundays and none of his staff were ever about when we went to pick up the diving gear. This was too much of a temptation for Brian and Gordon who would avail themselves of off-cuts of wood and the use of the machinery for their own ends. As far as I know, they never touched any of the new timber but anything that looked like it wasn't wanted was regarded as fair game. I would imagine that they probably held a "firewood order" just in case they were ever questioned. Gordon made use of this facility to construct a batch of 20 or 30 rabbit hutches which we popped on the lorry one Sunday morning and dropped off at his house at the end of the day. From then on, he raised rabbits in great numbers, always assuring his wife and children that he had no difficulty in find good homes for them whereas, in fact, he was fulfilling a weekly order from the local butcher. Gordon made sure that if we used any of the machinery, he cleaned it and sharpened it afterwards because Dennis Sharpe, "Sharpy", the man who looked after it had and eagle eye and would spot anything amiss.

Much later, after I had moved from Leicester to Bingham and was planning to build an extension on my house, I bought a load of heavy timber beams by "firewood order" from an old loading dock at Pye Bridge which was being demolished. One Sunday when Alan Lodge wasn't working with us, we contrived a light inspection programme and called first at Pye Bridge where we loaded the beams onto Brian's lorry and took them to Friargate Workshops. With Brian and Gordon's help, we put them through the big circular saw to reduce them to the size I wanted – some still a mighty 12" x 6" – but the trouble was that in our hurry to get the job done, we didn't manage to remove all the nails from the timbers. In my defence, I didn't realise what the consequences would be, but when we finished the teeth on the huge saw blade were badly damaged and although Gordon tried to re-sharpen them, there was no way we could hide the damage from "Sharpy". Fortunately, Brian came up with a solution and made sure he met up with Sharpy first thing on Monday morning before he saw the damage and offered him a "deal", the details of which were wisely kept from me but were on the lines of, "Sorry, Sharpy, but we used the big saw yesterday and it will need a bit of re-sharpening so would you like a load of manure for your garden." *OK* says Sharpy and from then on we are home and dry because by accepting Brian's deal, he couldn't then report us, even though the damage was much worse than he had been led to believe.

Just in case you think that we were total rogues, I would point out that what we were doing was no different to anyone else and management were perfectly aware that this sort of thing went on but adopted the approach of "what the eye doesn't see" so as long as we worked hard, completed all our tasks well and didn't overstep the (unwritten) mark. To some extent it was regarded as a "quid pro quo" because when the occasion demanded it, which it often did, we all went the extra mile and would work on late into the night if necessary to serve the needs of the railway. The consequences of anyone stepping over the line, however, could be severe and several railwaymen (some very senior) were lulled into thinking that they could get away with much more and paid the price. The most celebrated case was of a Divisional Engineer who used one of his workers as his private handyman and even got him to paint his house with railway paint. All went well until the "handyman" was made redundant when he got his own back on his boss by "spilling the beans". The engineer went to prison for that because once British Transport Police became involved, they (quite rightly) showed no mercy to railwaymen who did wrong.

Anyway, after any business at Friargate was finished, we would set off for the diving site. We only inspected bridges that could not be done without divers, so they were mainly the ones over canals and the large rivers that passed through the division including rivers Trent, Derwent, Ouse, Nene and Soar as well as smaller rivers and a number of special structures such as the Great Central line bridge over the Swithland reservoir. The justification for a structure being put on the diving list was that it was not possible for it to be examined by the regular bridge examiners wearing waders although later on, some examiners not wanting to get wet or muddy at all would declare that a bridge needed divers even though the water was very shallow. This meant that although some of the bridges required full diving apparatus, others simply required us to don our dry suits and put our heads into water that was around 1.2m deep. There were also some very long culverts where divers were used because of the protective suits that we wore and our training in "confined space" work. An example of these was a culvert in Burton-on-Trent.

Entry was at a headwall on the west side of the line, about a half-mile south of the station, next to a brewery site. The culvert started off with a diameter of 900mm but varied along its length and the air was full of the sweet sickly smells of brewing. On the first occasion that it was put on our list, we could not find the outlet on the surface, so Gordon was tasked with listening for our progress from the surface from any obvious points such as manholes. We set off and the first section went under the main lines but after a while there was a sharp left turn and we judged that the culvert was now following the railway towards the station on the east side of the line. We knew how far we had progressed because we were measuring the distance by tape measure but after a few more twists and turns we had lost out sense of direction. At each manhole we came to, we would shout but of Gordon there was no sign and although he denied it, I am sure he was sitting with the station supervisor having a cup of tea. Ever optimistic that we would eventually come to the end, we carried on following the ever-narrowing culvert until, at one manhole, we could hear road traffic passing nearby. Fortunately (and unusually) when we tried hard, we were able to move the manhole cover and, much to the surprise of the few people who were walking by, we emerged into the middle of the pavement on Station Road, several hundred yards beyond railway property!

Most of the other jobs during this period were fairly mundane, simply involving a "hands-on" inspection underwater and some basic measurements together with enough detail to produce a simple report at the end of the day so a typical Sunday would involve meeting up in Derby at about 8am (assuming Brian was on time), getting out on site as quickly as possible and completing the examination by around 1pm, leaving enough time for a pint and a "rub of the stones" (game of dominos) before the pubs shut and then back to Derby to clean and put away the gear before writing up the report. These various activities had an order of importance with the session in the pub at the top of the list followed by the exam then looking after the gear with writing the report very much last. On busy days, therefore, one of us took home the notes and did the report in his own time but whatever happened we always got a pub visit, usually somewhere near the site of the exam.

Once we all had a pint in our hands, we would ask the landlord for a set of dominoes and settled down to play our version a game called "Matador". This is a great game but totally incomprehensible to the casual onlooker because it is the only domino game that I know where the play does not involve "matching ends". Instead, the object is to make an "end" of seven so, for instance a 6 will go on a 1 and a 5 will go on a 2 and so on. Add in a few special rules about doubles and the three "Matador" dominoes which add up to seven themselves (the 6-1, the 5-2 and the 4-3) and we would soon have onlookers "saying you have played that wrong". Our response would be to just give away enough information to keep them interested but still puzzled and in the meantime up the tempo of the game and the loudness of the play while also throwing in random shouts such as "always keep a two-a-two", "old squareface" and "you've done me, you bastard".

Another job that we did during this period was a siphon located towards the top end of the Wirksworth branch, which was always paired with an examination of a long culvert in Wirksworth Station yard. One Sunday, Alan Lodge was on-call manager so couldn't be with us and I led the team. In this context, a "siphon" refers to a pipe (usually cast iron) that carries a stream under the railway where the line is in a cutting. The water enters at the upstream end and the pipe then slopes sharply down to pass under the railway before sloping back up to the surface on the other side of the line. The outlet is set a little lower than the inlet so the stream flows in and keeps the whole pipe full, but water gradually flows away at the downstream end. To carry out the examination, the diver must enter with full breathing gear, slide down the inlet slope then work his way under the tracks and back up the slope to the outlet. Great fun and comparatively easy but because of the shape of the siphon, any silt and debris entering will gather in the bottom, causing a blockage which the diver must clear to complete the survey, and this could be hazardous because if he became trapped, he would be unable to regain the surface. For this reason, we worked with a lifeline attached to a colleague on the surface. I absolutely loved the job and for some reason became the "siphon" specialist of the team.

Once we had done the siphon, we moved on the Wirksworth yard, and I entered the culvert at the upstream end. It was less that 750mm diameter and over a mile long, with many blockages and areas of collapsed brickwork. It seemed like hours till I emerged at the outlet, exhausted but really happy at two jobs well done. My colleagues had kept pace with me on the surface, lifting manhole covers where they could find them to keep in contact. As we were packing away the gear, in a merry mood and ready for a visit to the pub, Alan Lodge turned up in the yard to visit us during his Sunday rounds. Alan always seemed to have problems with parking and reversing (he once reversed off the edge of the dock in Holyhead) and after we had finished our chat and he was preparing to leave, Brian casually remarked, "I wonder if he will get out of the yard without hitting anything," whereupon, right on cue, Alan reversed straight into the large gatepost at the entry. We couldn't stop laughing and it may be the only time in my life that I actually fell down through laughter and lay on my back with my legs in the air impersonating a dying fly totally unable to get up for several seconds!

In October 1971, I moved into my first house in Bingham but diving work carried on the same. That winter was a hard one and at one stage there was major flooding in the area and one of our bridges – No. 92 on the Derby to Crewe line – was severely damaged by the fast-flowing River Dove. Unusually, the bridge carried a minor country road over the river, but it was BR's responsibility because it had been constructed by the railway company as part of the works associated with the original construction of the nearby line. The bridge consisted of a timber deck and beams supported on timber trestle piers and timber piles. The trestle in the middle of the river was badly damaged and partially collapsed and the diving team was mobilised to support the bridge section in making repairs. We opted for like-for-like repairs to the trestle first and this was done by me and the other divers cutting out the broken timbers, measuring the new ones and when they had been cut to size by Gordon working on the bank, we fitted them underwater using hand augers and large spanners to fit coach bolts and coach screws. The river was bitterly cold with water coming straight off the Derbyshire hills and our old dry suits offered poor protection.

Once the timber repairs had been completed which took a couple of days, we built a strong wall of concrete bagwork around the trestles which had the dual purpose of protecting the trestles from further damage and providing additional support in case the old timber piles were weakened under the bed level. There was some argument between the team on site about the correct way of making the concrete bags, but I was able to fall back on the notes that I made on my "Heavy Works" course at Watford, and I made sure the work was done properly. This involved using hessian sacks which were filled one-third of semi-dry concrete, properly mixed. The bags were then built into walls that were properly bonded in the same way as a brick wall. The concrete must have some water in the mix even though it is being placed underwater because otherwise it could just set as a crust of concrete and the inner material remain as loose cement and sand. Furthermore, the bags must not be overfilled because they must be pounded into position and lock in with the surrounding bags. As the concrete sets, there is a bleed of mortar through the hessian and each bag will bond to the next.

On the final day of the underwater work, we worked into the evening to finish the bagwork with flurries of snow around us and shallow pools of water next to the river already turned to ice. I got home about 9 o'clock in the evening and went straight to bed, exhausted, only to wake up a couple of hours later being violently sick and vomiting blood. We had only been in our house a few weeks and had not yet got around to registering with the local GP, but my wife rang him anyway (from a call box – we didn't have a home phone) and he came straight out and attended me. I just had a severe chill from all the cold water work but it was frightening for both of us. The team got some plaudits for that job, and I think it helped to persuade the "powers that be" to let us do some extra jobs away from the Division. As a postscript, the timber bridge was replaced some years later by a new concrete one. I took an interest in the job and asked the contractor what the condition of the concrete bagwork was like when they had to remove it. He swore and said, "It was like bell-metal – it took us weeks with breakers to remove it before we could put in the new concrete piles."

After this, my role with the team grew and I took on the responsibility of trying to expand the range of work. It would be sometime before we were able to work outside of our own area but eventually, with a more responsive Assistant Civil Engineer (works) in place, we were allowed to respond positively to a request from Crewe to examine structures on their area, including the famous Barmouth Viaduct. These structures had previously been examined by the team in its early days and the joke about that which was still circulating concerned a lack of knowledge of the geography of Wales. The team were booked into a guest house in Trevor which was only just over the border from England and in terms of travelling time, was further from Barmouth and most of the other sites than it was from Derby. When I asked Alan Lodge about this, he simply said that it was in Wales, so they thought it was bound to be near the bridges! Not falling into this trap, we found accommodation in a small B&B in Fairbourne which is just over the Morfa estuary from Barmouth. There were still some long drives to some of the bridges which were inevitably done at high speed, particularly on the way home for dinner and the pub (not necessarily in that order) but at least the biggest job – Barmouth Viaduct itself – was close by.

We were given a local Bridge Examiner to guide us to all the bridges and he travelled with us in the van. He didn't like the travel at all and after one particularly hairy journey he blurted out, "I have never kissed my wife for 20 years but this week, I have kissed her every morning before I set off for work just in case I never see her again." Barmouth viaduct is a long structure but much of it is dry at low tide and was examined by the regular bridge examiners. We concentrated on the section near the north shore which

212

comprised a group of four metal cylinders in about 50 feet of water. These supported a wrought iron superstructure that once had been able to be swung for tall ships to pass through, but which was now fixed in the closed position.

The water ran fast through there between tides but had reasonably long periods of slack water at high and low tide which enabled us to do the exam. The water was crystal clear and was such a change from the muddy rivers of the Midlands that after completing the exam, we stayed on just to enjoy the diving, chase some fish and collect mussels. We all took bags of fresh mussels home, but I really didn't know what to do with mine, so they sadly went to waste. The food at the digs was excellent and the lady looked after us well. Her father lived with the family, and he had never learned to speak English. He was interested in what we were doing but had to get his daughter to translate for us. There was no pub in Fairbourne, so after the evening meal we would drive the van the short distance to Morfa Mawddach Station and then walk over the bridge to Barmouth. It was summer and the weather was pretty good to us, so it was a really enjoyable excursion. On the way home after a few pints, it was always necessary to empty out bladders as we crossed the bridge, and it became a tradition to piss into the waters of the estuary below. One particularly beautiful clear night as we stood in a line pissing off the bridge, Brian Garner issued the immortal words "just look at that conster-fucking-lation" which have remained as a part of the team folklore ever since. We even managed to convince a new member of the team, Paul Barnes, that the correct grammatical term for what was said was a split infinitive!

After our trip to Wales, we were keen to go further afield but opportunities were kept to a minimum until I was on a course at Watford again and got talking in the bar to Peter, an engineer from "Sealink" who was responsible for several harbours. In 1970, "Sealink" had been formed as a wholly owned subsidiary of the British Railways Board to take over the running of all its ferry services and ports. Once established, it operated with only top-level support from the main British Rail family and established its own structure of engineers and managers. Later on, "Sealink|", being by then semi-independent of the BRB was one of the first parts of the business to be sold into the private sector. Anyway, Peter and I talked about harbours, and I mentioned my job as a diver and how we would love the chance to examine one of his harbours. Peter was very enthusiastic and said that the harbour at Holyhead had not been examined for some years. The work had always been done by members of the port's own workforce based in accommodation on the harbour and he still had a couple of his team who were trained in the use of "standard" diving equipment. However, some years before, the old standard equipment had been pensioned-off and all new SCUBA equipment had been purchased for his divers, but they had declined to use it. He sympathised with them because they were already over 60 years old and had no wish to learn new ways. I am not sure if any major effort was made to persuade some of the younger men to do the job, but if there was it had been unsuccessful.

The port was heavily unionised, and Peter was reluctant to cause trouble by bringing in contract divers, so things were at an impasse. We were his solution and if we would take the job on, he promised that all the unused diving gear in his stores would be transferred to our team. I explained that if I raised the matter back at Nottingham, there would almost certainly be objections so it was agreed that he would make formal application through the Chief Civil Engineer at the BRB. So it was that some weeks later, Alan Lodge was called into the Engineers Office and that the expertise of our diving team had been recognised "at high level" and we were to prepare ourselves to examine Holyhead Harbour but that it mustn't interfere with our normal duties. After some discussion, it was agreed that we would be released on a Friday lunchtime to set off for Holyhead, we would dive on Saturday and Sunday then return home on Sunday evening, but we had to be back at work on time on Monday morning. The first year we did, I think, 3 successive weekends like that which set a pattern for several years to come, until "Sealink" privatisation took place and we lost the work.

Alan Lodge and I drove over to Holyhead to meet up with the local staff and do a recce of the job. We wanted a full day and arranged to be there at 9am so I picked up Alan at home in Derby at about 6 o'clock and we went flat out all the way to Holyhead, arriving just in time. Peter was still busy with other port issues, so we were introduced to the local works supervisor and former diver whose name, of course, was Mr Jones. We waited for him while he dealt with assistant supervisor who had incurred his wrath. Mr Jones's first language was Welsh and he was berating his underling in that language which, it seems to me is ideal for giving someone a good bollocking. It was fascinating to hear the stream of Welsh invective but of course, Welsh is an old language and not all modern English words have an easy translation so amid a long flow of Welsh, the English words "fucking shackle" suddenly appeared. Both Alan and I burst out laughing and fortunately, our Welsh colleagues also saw the funny side.

There were occasions when we worked in Wales that the locals were less than friendly to us and once, when we went looking for a "pie and a pint" after a long day's work in a village in Mid-Wales, the reception was so unfriendly in the local pub that we turned tail and left. Fortunately, this did not apply to the port staff in Holyhead who were a mixture of Welsh, Irish and English with a heavy "Scouse" accent and all of which were friendly and helpful once they saw that we could do the job. Mr Jones showed us all the SCUBA equipment that was to become ours and it was very good with large "twin set" air cylinder that would give us longer time underwater. He also accompanied us around the harbour and along with Peter, showed us the areas that they required examining. For a team that up till now had mainly been involved in shallow water inland inspections, task was quite daunting but I, for one, relished it and we were all determined to do a good job. Inspections were urgently required of the main ferry berths; the container ship berths, and an area known as the "Cattle Dock" which was now disused since the movement of live cattle from Ireland by ship and train had ceased but the dock wall supported various roadways within the port. Alan and I returned to explain the work to the rest of team, and we made arrangements for our first trip.

We didn't want to take a heavy van load of equipment to Holyhead or to tow a boat trailer all the way from Derby, so we devised a plan to send our gear by rail and travel ourselves in a couple of cars. We also booked "digs" in a Guest House close to the harbour. We duly loaded up a box van in Sandiacre yard with all our equipment carefully stowed and with the added luxury of a couple of chairs and a camping stove and labelled it up to Holyhead Station Platform 3 and put it into traffic to go there by goods train. Mr Jones informed the station supervisor that it was on its way and when we arrived a few days later, it was parked on the buffer stops at the end of Platform 3, as requested. We just had to open the doors and we had a ready-made changing room, kitchen, equipment store and drying room. The old platform 3 at Holyhead was not often used in those days so we got used to a degree of privacy until one day when a party of loud Americans came walking by on their way from the Dun Loaghaire ferry to the London boat train. We were caught getting changed out of our diving gear while Gordon was rustling up some bacon sandwiches on the camping stove. One large lady said, "Gee, boys, I see you are eating Irish rashers." She had clearly been impressed with her Irish breakfasts but had forgotten that she was now in Wales (or perhaps she had the usual American knowledge of geography and didn't know there was any difference).

Getting caught with our pants down was a bit of a theme with the diving team, inevitably really since we often had to change in the open air, but my own experience was in the B&B at Holyhead. The landlady, who we confidently called Barbara, was excellent and every day after we got in from work, she would knock on the door of our room with a tray of tea and cakes. On one occasion, most of the lads were in my room for some reason and I was stark naked, just about to get dressed when she knocked on the door and said, "Are you all decent?"

Quick as a flash, Brian shouted, "Yes come in, Barbara," and all I could do was drop into one of the chairs and hold a towel across my lap. Many years later, Brian got his "comeuppance" on a filthy dark day down on the Southern Region when a woman police officer shone her torch into the van to see what was going on beside the railway and caught a naked Brian in the act of getting changed. We let him mumble over the answers to her questions for some minutes before we waded in to explain the situation, just in time before he got carted off for being a pervert. While talking about Barbara I must mention the last time we stayed with her. We knew it would be the last because "Sealink" was being privatised and we would visit Holyhead no more, so we all bade our final farewells and she said how much she had enjoyed having us stay there but had just one thing to point out: "My name's not Barbara," she said.

The diving at Holyhead was both exciting and exhausting with diving going full blast until dark followed by drinking down the pub and a fish supper which, unfortunately, was possibly the worst fish and chips I have ever tasted. We always referred to this as Irish fish though whether it was or not, I have no idea, but it certainly wasn't local. For some reason (possible due to alcoholic intake) we contrived on one occasion to each eat three portions during the course of one evening. To fully examine the harbour walls and docking facilities, it was necessary to change depth frequently and the constant changes of pressure are quite debilitating to a diver, causing a condition which we all referred to as the "turbles" (as in turbulence), a general feeling of light-headedness and malaise that could only be cured by pints of ale. Nevertheless, we enjoyed it all and took plenty of underwater photos which enhanced our reports. We also found time to pick up a lobster or two which we would find hidden in the crevices of the harbour walls. Gordon was on one occasion despatched to one of freightliner container ships where he persuaded the cook to let him use the stove to cook the lobster. It made a change from the usual fish and chips and it gave us the idea that maybe we should try some other food so one evening we drove across Anglesey to the town of Almach where we proposed to have a curry. Geoff Barker was very reluctant and left early in the meal. When we met up with him later and asked him what was wrong, he replied that he didn't want any more of that "Dogshit Vindaloo" and so coined another phrase that stayed with the divers for evermore and was used to describe any meal that we didn't like.

On one trip, as well as the harbour structures, we had to drive out along the breakwater, which was over a mile long and examine the end of the breakwater itself, which was the only part that was not exposed at low tide. We devised a plan, using ropes to hang onto because of the extreme current and with great difficulty worked our way around the structure, hanging on the rope and examining as we went. I think it was Geoff Barker who was examining and Alan Lodge was on the safety rope. Geoff was getting further and further away and when the end of the rope came to Alan, he shouted for another one to tie on because "he is skating it". The rest of us were not convinced and ignored Alan's instruction and pulled Geoff back in. Poor Geoff had been desperately trying to get back down the line but as fast as he pulled on it, Alan gave him more rope. Unfortunately, the general turbulence of the water had made any attempt to pass signals but giving "pulls" on the rope impossible. It was generally agreed that if he had enough rope, Alan would have let him go all the way to Ireland before he pulled him back.

Of course, all these weekends at Holyhead were difficult for my family life and so, on one occasion when we were diving there on 2 successive weekends in the summer, I booked a holiday for the week and travelled over to Holyhead by train on a Friday evening, taking my wife and 2 sons with me. Ina and the boys got accommodation in a different B&B just up the road whilst I joined the lads at Barbara's. Over both weekends, while I was diving, Ina brought the lads down to spend time with us while we dived. I was particularly proud to show off my skills to my boys as any father would be, but they eventually tired of just watching and needed some other entertainment, so Ina took them to Saturday afternoon pictures where she endured fighting and rioting by the local kids and a good number of threats by the manager before they settled down to watch the film. On the Sunday evening, we caught a train to Llandudno and trailed the streets to find a B&B to stay at for the week. It was 9 o'clock at night before we found one and I think Ina was envisaging us all sleeping under the pier but the one we did eventually find was very good and we all had an excellent week's holiday before going back to Holyhead on the Saturday morning and repeating the procedure the following weekend.

On one occasion when we were at Holyhead, Mr Jones showed us the equipment that he and the other "standard" divers had used which included old dry suits and a hand pump but the items that interested us most were the helmets. We were told that they were about to be sold as scrap but if we wished to purchase them, as serving railwaymen, we could do so. There were, I think, 4 helmets and Geoff, Brian, Alan Lodge and I all got busy with the "firewood orders" and became the proud owners of the magnificent helmets. Working diving helmets are made of copper but are "tinned" to provide a corrosion resistant surface. I was all for sanding off the "tinning" to give a shiny copper appearance, but Brian wisely warned me off this because it would have reduced the value of the helmet. Mr Jones also told us a strange story about one of his colleagues who, he claimed, had jammed his finger in a crevice and was unable to free it. To return to the surface, I was told, he cut off his own finger. I must confess that I rather treated this gruesome story as a typical "divers' tale" and took it with a pinch of salt but many years later I was to find out the truth.

The BBC programme "Antiques Roadshow" was visiting Southwell Minster which is only a few miles from where I live, and I went along with my helmet. The presenters of the show took some interest, and I heard Hilary Kaye and Paul Atterbury discussing the helmet and what they knew about it. It transpired that they wanted to put the helmet into the show because they had not had one before and thought it would something different but neither of them knew much about diving equipment. I was asked if I could come back in the afternoon to be interviewed for television and of course, I readily agreed. When I returned, I was taken straight to makeup and then wheeled out to meet Hilary at a table set up in the grounds of the minster. Hilary was quite "flirty" with me, a technique clearly designed to put me at my ease in front of the cameras. That approach succeeded with me, and I was soon launching into the tale of diver cutting off his own finger with the caveat that although I knew that this diver had used my helmet, I had no idea if the story was true. Hilary's researchers had done their job, and she was fully conversant with the helmet and was able to tell me that it dated from the 1950s, that it was not particularly rare or valuable but that they were in demand by collectors and interior designers who would put them into nautically themed restaurants so I could expect to get around £3000 pounds at auction.

A few weeks later, an email was forwarded to me by the BBC which told the full story of the "finger" incident. The email came from the master of the cruise ship "The Saga Rose" and he told me the full story of his father, who was the diver who got his finger stuck. His father was a merchant seaman in WW2 and was a member of the crew of one of the ships in "operation pedestal", the famous convoy sent to relieve the island of Malta. The convoy was heavily bombed and the ship on which he was sailing was sunk but he was rescued from the sea by a Royal Navy vessel. He was then transferred to the crew of the SS OHIO, an oil tanker that was carrying fuel for the fighter squadrons based on the island. The OHIO itself was hit several times but eventually limped into Valetta Harbour with naval vessels secured to each side to help keep her afloat. The whole crew were treated as heroes, as indeed they were, and after the war a film was made of the whole "Malta Story".

Our hero left the merchant navy at the end of the war and got a job as a diver and general worker with the newly nationalised British Railways at Holyhead harbour. Sometime later, he was diving in the harbour and carrying out some work on one of the harbour walls when his hand became caught in a crevice in the wall and because of the stiffness of the diving gloves, he was unable to get it free. He tried for some time but remained trapped and to make matters worse, his telephone communication failed so he could not relay his predicament to his colleagues on the surface. Eventually, in desperation, he signalled on his lifeline to be hauled up to the surface, hoping that his hand would be freed as they pulled him up. His hand did become free, but a finger remained behind. When he got to the surface, they despatched him to the local hospital where he got a roasting from the matron for dripping dirty seawater and blood on her nice clean floor.

All the trips to Holyhead had been in the summer but one January day I took a call from Peter Hunt to say that a section of the harbour wall had collapsed and could we get to Holyhead urgently to inspect the underwater section. Of course, we jumped at the chance and the next day we were on our way. When we got there, we were met by Peter and several other senior managers from "Sealink" who were most anxious about the situation because it was a part of the old cattle dock wall that had collapsed and although cattle were no longer imported, the dock was important for the operation of the harbour as a secure place where small repairs could be done to the ships and where they could safely be moored during stormy weather. We saw about 10m of wall had collapsed into the sea, taking some of the fill with it but also leaving a large section of concrete paving overhanging the void. Brian and I got dressed and launched the inflatable together with Gordon who acting as boatman. I think Colin Tigg might also have been with us but if he was, he would certainly have been up top "my-chappying", which is what we called the act of talking to bystanders while someone else did the work. We got into the water and carried out a successful inspection dive and then we surfaced at the base of the wall.

The group up top were anxious to hear our initial report but as we shouted the results up to them, there was an ominous crack, and a large section of the overhang broke off and a 2-ton lump of concrete came plummeting towards us. It missed Brian by inches but as I pushed away from the face of the dock in an attempt to avoid it, I was struck on the shoulder and carried down by the concrete to the bed of the harbour. Fortunately, because I was floating, the force of the slab was not as hard as it might have been and by the time we reached the bed, much of the force had been dissipated, but I did not have my mouthpiece in or my mask on so the first thing I did was to put my mouthpiece in and breathe deeply. After that I rested for a while and recovered my composure, then I spent a while searching for my mask before giving it up and slowly returning to the surface. Of course, all this had taken some time and all the others feared that I might have been badly injured or trapped under the debris. Brian had already completed a short search for me and returned to the surface with a negative report when I surface way out in the middle of the harbour to the relief of all concerned. The reaction of the divers in the party was predictable and I was berated for losing the mask, but they had clearly been troubled by the incident. At the time, it didn't bother me particularly and we all went on the complete the work but looking back now, it was one of the very few occasions in my life that I had been in serious danger.

After we lost the Holyhead work, we were restricted to local jobs for a while but during this period, we did start to get better equipment including new valves and all the diving equipment from Holyhead, but the main bone of contention was the old 2-piece dry suits that were difficult to get on and off, uncomfortable to wear and constantly leaking. In the meantime, we saw pictures of amateur divers using new wet suits which, looking a much better option and after much lobbying, we were each measured up and issued with new wet suit. The principle of a wet suit is that once the cold water has been heated by the body, the neoprene suit will retain a warm layer of water to protect the diver from cold. They were certainly much easier to get on and I was convinced that they would be warm to wear in all conditions.

The first job that we used them on was the River Amber at Ambergate in the middle of winter and a layer of thin ice covering the water. I quickly donned my wet suit but noticed that Brian and the others were hanging back. Brian was not convinced about the warmth of a wet suit so they all watched me stride confidently into the water. I screamed in pain as the cold water hit me and I can honestly say that I have never felt so cold in my life, but I had to carry on and do the exam otherwise I would have lost considerable "face". I found out that whilst a wet suit will quickly warm up in deep water, give a bit of vigorous activity, when striding about in shallow water, the suit never really holds the water long enough to warm up. Eventually, we got much better wet suits which did keep us warm during summer conditions, but we had to put up with a lot of cold dives until much later, we received new, modern one-piece dry suits.

Gordon Issott (left) and Brian Garner in the original dry suits and SCUBA full face masks

My Helmet

Barmouth Bridge

Ladybay bridge, Nottingham

Chapter 20

Diving – I'm in Charge

The diving team continued to run under Alan Lodge's leadership throughout the mid-1970s but the personnel gradually changed. Colin Tigg had already joined the team but over that period, Alan Lodge stopped diving and eventually retired to Spain and Geoff Barker left the railway industry for a time and so was no longer available. New blood was required, and we recruited Paul Barnes who was a navy-trained diver. He was working as a trackman with the P-Way department at the time, but we were able to arrange his transfer to the technical office and for him to go on a training programme for his new role. From the technical office, we recruited Graham Williams and Pino DeRosa who I have mentioned before and who was, by then, working for "spanner man" in the building section. Pino elected to join the team despite intense opposition from "spanner man" who told Pino that we were a bunch of rogues who would ruin his career – how wrong he was! Finally, a promising "youngster", Nigel Stockdale, who had moved down from Cumbria to study civil engineering in Nottingham was also recruited.

When Alan Lodge stopped diving, I was put in charge of the team, mainly, I think, due to the fact that I was already working in the bridge section which was the natural "home" for the team leader but also, possibly because I was the most enthusiastic proponent of the diving team and I had made it clear that I saw possibilities for its expansion. Local diving followed its usual pattern with Sunday morning dives followed by a pint or two. The long ride home after those jobs could cause some embarrassment and one Sunday, Brian was taking me home when we both decided that we needed to wet the rear wheel of his lorry. We pulled into a lay-by which had been formed from a loop of tarmac that remained after the road had been straightened and had a convenient screen of trees. Unfortunately, the road had a slope and as we made our way back to the cab, we saw a stream of steaming yellow liquid heading straight for a family who were having a picnic on the roadside grass. We were off and away before we got a chance to see if it reached them on not. The diving team had always had a couple of "Diver's Dinners" each year where we took our wives to a nice restaurant. These were normally held on a Friday evening so as not to affect our Sunday diving operations but on one occasion, it had to be on a Saturday and since I had transport arranged to the venue, I proceeded to have rather a lot to drink.

Next morning, when Brian picked me up in his lorry, I was feeling the worse for wear but, determined not to let it show, I was first in the water which happened to be a rather pleasant dive in the River Trent on the outskirts of Newark. When I dived, however, the combined effects of a tight diving suit, heavy equipment and the pressure of the water had its effect on me, and I had a sudden need to vomit. I whipped the mouthpiece out of my mouth and let it all go, watching with fascination as it floated around me and followed me to the surface. While the team was still under Alan's leadership, we had a few interesting "away" dives, two of which were associated with British Rail Engineering (BREL) premises.

The first was at Doncaster Plant and was unique in my experience. The Plant had a tall water tower to supply non-drinking water to all its facilities and parts of the station and each year during a two-week summer holiday shutdown, the tank was drained and inspected. On the previous year, some defects had been noted in the metalwork of the tank but there had not been enough time for new sections to be manufactured and fitted and it was feared that the same would happen the next year unless the tank could be inspected some weeks before shutdown to allow time for new parts to be made. That was where we came in and one Sunday, we found ourselves being met at the gates of the plant and escorted to the steps at the bottom of the tank. We then had to climb stairs and vertical ladders to the top of the tank (see photo) with all our diving equipment. Once that was out of the way, the job was a joy to do since the water was perfectly clear but took a long time as a "Braithwaite" tank (which this was) comprised metal panels that are supported internally by a mass of horizontal and diagonal tie rods, all of which had to be examined and measured. The second BREL job was to examine a reservoir situated alongside the railway several miles outside Crewe. The reservoir was connected to Crewe Works and to the town itself and for many years, it had not only been the main water supply for all the railway premises in Crewe but also the town which had largely been built by the railway company because, like Swindon, Crewe was a real "railway town".

Our job was to inspect and survey the bed of the reservoir which was about 3m deep and the slope of the earth dam. Again, it was a pleasant job in clear water and to ensure that we examined every part of the bed, we tied long ropes together to divide the area into manageable sections and only moved on to the next section when the first had been completed and documented. Geoff Barker was a very keen rope man and prided himself and his care of the ropes and his expertise with knots. I had been known to upset him by not taking this job seriously and once, in the changing wagon at Holyhead, I had infuriated him by trampling all over his ropes when I was dripping wet from a long dive in the harbour. I therefore decided to get my own back and while I was diving one of the sections of the reservoir, I quietly undid the knot linking two pieces of rope together so that when time came to move it up and the attendants pulled on the line from each end, it simply came away in two parts. I thought it a great joke, but poor Geoff was mortified and kept saying, "I don't know how it could have happened." Later, I realised that what I had done was very silly and that it had really upset Geoff and to this day, I regret doing it.

As lunchtime approached, we realised that we would not be finished in time for a pint, so we decided to send Gordon to the very posh country pub that was just down the road with a couple of large water carriers and instructions for him to come back with them full of draught beer. Unfortunately, the landlord would not serve him at first and only did so after Gordon explained the situation. He then reluctantly agreed, but as Gordon left, he could be heard tut-tutting with his cronies about the feckless labour employed by British Rail.

During this period, we did have one very interesting "home" job on the River Trent. Just South of Trent Junction, where the Midland Main Line crosses the River Trent, there are 2 large bridges, 27 and 27A, which carry the Main Lines and the Goods Lines over the river. When the line was built, the railway company had to move a weir some 100m downstream and that weir remained in British Rail ownership and was therefore one of the structures that we regularly examined. There was concern about the condition of the weir and about the bed erosion that was happening around the bridges, so a scheme was developed by Mike George, one of the Senior Technical staff in the bridge section, to repair the weir and to lay "Reno mattresses" across the bed of the river in the vicinity of the bridge. Reno mattresses are constructed of wire boxes filled with stones that are fitted together on the bed of the river and offer good protection against scour. A contract was let for this work as it was well out of the scope of our own staff, but the diving team were required to visit the site to check that the work was being carried out correctly. As work progressed it became clear that a permanent presence of a railway diver was required on site because it was essential to confirm that each part of the Reno Mattress was correctly placed before another was laid.

Furthermore, we had a good relationship with the contractor's diver (who was also their site agent) and rather than him hiring in additional contract divers (the cost of which would have been borne by the railway) to place the mattresses, just for our team to go down and check them, it was agreed that we would carry out the work ourselves. The contractors asked for permission to pay us direct for our work but of course, this could not be allowed by BR, however, it was approved that if the work was completed satisfactorily, we could attend a celebratory event and accept a small gift.

The work went well, and I really enjoyed actual underwater work (as opposed to examinations). The mattresses would be lowered onto the bed of the river by crane, and we would ensure their correct positioning before undoing the shackles so the carrying frame could be lifted clear. We then went round the edges of the mattress and fixed it to the adjacent ones with wire. It was all great fun, as was the assistance that we provided to the weir repairs. The southern apron of the weir had to have its concrete surface replaced and we assisted with the placing of temporary barriers to divert the water to the centre and north side. The concrete was then pumped over to the apron and vibrated into position. While this was going on, a bung that had been placed in a hole in the sheet piling became loose and threatened the whole operation, so I had to dive down through about 1m of wet concrete to re-fix it. No great problem but no visibility of course and all my equipment had to be carefully cleaned afterwards, otherwise, it would have been ruined. At the end of the job, which was completed on time (very important because if not finished by October, river levels would mean the site being abandoned for the winter) and on budget, the contractors did us proud by hosting an evening meal at a popular "carvery" in a Nottingham Hotel and providing accommodation for those who needed to stay over. They also provided the gift of a silver mermaid on a chain for all the wives and presented the divers with good diving watches. I still have my watch and wear it proudly and my wife still wears her mermaid from time to time. The whole job was an example of how clients and contractors can work together for the mutual benefit of both parties and that rewards can be given to people who have worked well without any suggestion of impropriety.

At the time, British Rail had three diving teams carrying out inspections and all with a very similar set-up in that they used civil engineering technical staff who were trained in SCUBA diving. One, based in Glasgow, covered the Scottish Region and one covered the Southern Region while our own team just covered Nottingham Division but not the whole of the London Midland. Other Midland Divisions used contract divers and I know that the engineers at Nottingham were reluctant to let us expand into other Midland areas. All of the teams relied on enthusiastic leaders, like Alan Lodge and myself to continue operating and one by one the others went into decline. New regulations for commercial diving, introduced in 1981, also made it more difficult for part-time teams to operate. The first to stop was the Southern Team and the former team leader contacted me to see if we would take over at least some of their work. I was excited by the prospect, but I knew that I would have to tread carefully because if the managers at Nottingham got to know the extent of the commitment that I was about to take on, they would certainly veto the idea. I decided that we would just do the work and say very little about it to anyone. As long as we could keep the work to weekends only and also complete our own area schedule, then we would probably be allowed to get on with it and that is exactly what we did.

The plan was that on our first trip we would support the existing team leader in carrying out urgently needed inspections of the ferry berths at Folkestone Harbour and the boat train dock at Dover. I am afraid that I can't remember the name of the man from the Southern Region, but I can remember that he was a good enough diver but that he was a typical London wide-boy whose main preoccupation appeared to be to do as little work as possible but to spend a long time telling everyone how difficult the work was and how good he was at doing it.

He met us on the Friday evening when we arrived at Folkestone and over a few pints, he outlined the arrangements for the work the next day. The plan was to meet first thing in the morning and carry out the inspection all day but by the time he arrived, we had already gone through the harbour security (who were aware that we were coming) and had set up camp in the harbour. When he arrived, he spent a few minutes with us but to be honest, he needn't have bothered because it was clear what was needed so we just posted Gordon at the seaward end of the harbour wall with instructions to advise us of when boats were approaching so that we could be recalled to a place of safety while docking took place. It all worked very well, though I doubt if such an informal arrangement would be allowed today. The dives were all interesting and revealed several areas where repairs were required as well as views of a great deal of sea life.

The next day we travelled to Dover where the arrangements were that I would support the Southern Region diver on an examination of the train dock gate. At Dover at the time, the train ferries that carried passenger coaches and freight wagons had to enter through a lock gate that would be closed behind them and then the water level in the dock would be regulated so the level of the rails on the boat matched closely to the level of the rails on the dock. In the open position, the dock gates lay on the seabed within a concrete housing and were raised and lowered by hydraulic rams. In that position, there was a space of less than 1m between the gate and the housing through which we had to dive to inspect the gate, the rams and the bearings. Of course, this meant that we were under the gate for a long period with no ability to rise to the surface in case of emergency, which a bit disconcerting. It was one of the few occasions when I would have preferred to have been working with an airline to the surface.

In the end, the week's work was completed, and we submitted our reports which were deemed satisfactory. The Southern Region man did not come on any more dives with us, but he remained our contact for some years as we continued to expand our Southern Region work. We had a couple more visits to Folkestone before we lost the harbour work but never returned to Dover. It was at Folkestone on one summer day that Gordon announced that he had found a cabin with a working fridge in it, and he had put in some beer to cool (it was mid-summer). We worked and sweated all day, looking forward to a cool beer upon completion of the work but when it came, the beers were frozen solid because the only compartment of the fridge that worked was the freezer!

One of the main sites that we visited over the years was Rochester, where the railway crosses the Medway on a mighty bridge which has three sections, one for rail, one a former rail bridge that now carries the coast-bound road carriageway, and one more modern section that carries the London-bound carriageway. Over the years we have probably examined this bridge over a dozen times and there are more "diver's tales" associated with it than any other structure. The first time that we inspected it, we were still sending our equipment ahead by rail to provide us with a base for using the compressor to re-charge our air cylinders and for somewhere to get changed before going back to the "digs". Brian and Gordon set off from Nottingham Friday lunchtime by road and went to the wagon which was stabled in Rochester Goods yard. The rest of the team travelled down by train after work and arrived early evening. When we found Gordon in the goods yard he was on his own and told us that Brian had taken a walk down the cess to look at the bridge and get an idea of what we would be facing the next day. I thought, great, no problem and then I thought, I wonder if Brian has ever worked on live third-rail tracks before? It was only our second visit to the Southern and the work at Folkestone and Dover had not necessitated crossing live lines so I suddenly realised that we might have a big problem. Fortunately, at that moment, Brian came strolling back and I ran across to him and said, "Do you know what that is?" pointing to the third rail.

Now, Brian can be a great one for the "wind-up" but on this occasion, I am sure he had no idea that stepping on the rail would, at best, give him a nasty burn and at worst could kill him! Of course, Health and Safety was less than perfect in those days and information such as this relied on the experienced person looking after others whereas before long, such access to the track would not take place without a full safety briefing. We had arranged with the Medway River Authority for permission to dive on the structure and for the hire of their workboat, the "Medway Otter" and next morning we met it at Strood pier where we were introduced to the 2-man crew and loaded our gear. The first examination took a whole weekend because we had a lot to learn about the structure and the riverbed at this location, but the work was made easier by the help of the Medway lads who, as well as navigating their boat superbly, they also allowed us the use of their tender. The bulk of the diving was on the central pier, and this could only be carried out on slack tide but it was a moot point whether it was easier to work on high tide (deeper but clearer water) or low tide (less depth but poor visibility). The outer piers and the abutments were generally shallow at low water and some parts even dried out completely.

We did some initial work on the centre pier at low water, enough for us to know that we were dealing with a very complex structure. The pier had been built with two large metal caissons supporting cast iron beams that in turn supported the masonry pier. In addition, there were wrought iron piles at the connection with the road bridge pier and a large circular pile supporting the downstream cutwater. Without any detailed plans of the structure and working in complete darkness with silt-laden water that just reflected the beams of our lamps, understanding the details of the structure was very difficult and in truth, it was not before we had visited the bridge on two further occasions, obtained copies of drawings of the bridge from the railway archives in Waterloo and details of a paper given to the Institute of Civil Engineers by the designer of bridge that we fully understood its construction. What was very concerning at first was the large amount of concrete bagwork surrounding the pier that was in very poor condition and crumbled away when we touched it. It took us both high and low tide inspections to be sure that this bagwork had been added some time after the construction of the pier and was contributing nothing to its strength. We speculated that early diving inspections had revealed what appeared to be a void in the pier, but which was just the original shape where the rectangular caisson met the round nose pile and that the divers had decided to fill the gap "to be on the safe side".

Gordon was the attendant on these dives, having stopped diving with ear problems a few years before and we relied on him to fetch and carry. The tender to the Medway Otter was a small wooden dinghy that was propelled by a single oar at the stern of the boat. Gordon was instructed to take the tender to fetch some sandwiches from the van that was parked at Strood Pier. Now, propelling a boat with a single oar is a real skill that Gordon did not possess. However, he tried it at first with the boat still tied to the "Otter" and seemed to have mastered the technique, so we cast him off into the main flow of the river only to see him swept away towards the sea with no control whatsoever on the boat. It took about 30 minutes to chase him down and bring him back to the bridge! At the end of the day, the Medway Otter dropped us off at Strood Pier, along with all our equipment which then had to be carried up the walkway to the van and then transported back to Rochester goods yard. By the time evening came we were exhausted but still managed to consume a few drinks.

On subsequent years, we dispensed with the railway van but still used the "Medway Otter". The next examination went well and we finished the main diving on the Saturday in early evening with another site to be dived on the next day, we had to set up our compressor on Strood Pier and after all the gear was unloaded from the boat we began the arduous (and noisy) task of charging the air cylinders and carrying all the equipment up to the road van and loading it for use the next day. We had decided that we would all work together on this task until it was complete the go back to the digs and change before going out for a few pints, however, the lights from the Riverside Tavern, just downstream of the pier proved to be too much of a distraction. First, it was just Brian and I who disappeared into the pub and sat drinking a couple of pints on the terrace while we watched the rest of the team still working away. As we watched, it became clear that others had spotted that we were missing and knew we must be in the pub so, one by one, they all followed us to the pub where we watched "poor old Gordon" as he was frequently referred to continue to load the van on his own while occasionally looking up to see where we all had gone. Eventually we took pity on him and with 2 of us making sure we didn't lose our table as the pub filled up, we helped him with the rest of the loading. We never did leave the pub to get changed that evening and we formed a somewhat smelly corner of the room where later on a very good music session took place.

We returned to Rochester many times and because of our detailed knowledge, we continued to do the inspections even after privatisation when the team was operating as a part of Bridgeway Consulting. The first occasion was soon after the company had been formed and we were determined not to lose the work so put in a rather low price which did not allow for hiring the Medway Otter. Instead, I contacted my Uncle Clifford who lived in the area and kept a cabin cruiser on the Medway. The job went well and Clifford, even though he was over 70 years old at the time, did a great job and had a whale of a time. He said later that he had always wanted to do some work with his boat rather than just the occasional pleasure cruise. Of course, we took him for a few pints afterwards and I was amazed how well he fitted in with the team.

During the course of the work, The Medway Otter paid us a visit and the crew were clearly displeased that another boat was working on their patch but we had all the necessary paperwork so there was little that they could do. The next time we came down, some years later, we hired the Medway Otter again and reinstated the good relationship that we had with them. This last occasion was the biggest job that we did, and we were contracted to provide a full bed survey as well as the usual inspection and we also had to produce detailed drawings and underwater photos of the central pier. To do the survey, we brought our own survey team with us, and they followed the movements of the survey boat with their digital technology so that we could ensure an accurate profile of the bed was produced. Using our underwater camera and working on high tide when the water was clearest, I was able to photograph patterns of small cracks in the cast iron beams, giving much more detail that we had previously been able to provide. The visit did not start too well, however, because the first day of diving was a spring tide which meant that low water was VERY low. When we first entered the water, I rolled backwards off the boat adjacent to one of the side piers, only to land on my back in water that was only 150mm deep! Of course, everyone on the job, including the crew of the Medway Otter, saw the incident, much to my embarrassment. Once I got over that, I realised that all the inspections, except for the central pier, could be done on foot and without SCUBA gear which made that part of the job easier. We also inspected much of the central pier at low water when features previously only seen when diving was revealed and I have to say that we were able to confirm the accuracy of previous reports.

We spent many years doing inspections down on the Southern Region and the range of structures examined was extensive. Near to Rochester is the

Kingsferry Bridge, a lift bridge that carries both rail and road over the Swale to Sheerness. The interesting thing about this was that the big concrete piers which appear solid at water level are, in fact hollow, so on the first dive, I went down under the water, expecting to follow a concrete wall to the bed, only to find myself in a chamber and when I tried to surface, I found that I was bumping my head on a roof as I had inadvertently strayed into the chamber. Once that happens, it is easy to become disorientated so I had to follow the roof very carefully to the point where I could see clear daylight and a way out.

Another centre for inspections was Lymington and we found very good digs there, run by a lovely elderly lady, right on the harbour. It was a good centre for inspections along the south coast from Portsmouth through to Bournemouth so we often used it, but the last time we tried to book in, the phone was answered by a man who said that the B&B was now closed. When we told him who we were, he said he was very sorry, but his mother had been ill and was no longer running a B&B, but she often mentioned us and the pleasure that she got from our visits, even if we were a bit rowdy.

For many years, we booked our accommodation ourselves when working away and then submitted a claim for expenses that was paid at a standard rate. Of course, this meant that if we used low-cost accommodation, we would be able to make a little extra money although we sometimes took this to extremes and "caught a cold". That was the case on our first trip to central London to examine the rail bridges over the Thames. I booked accommodation in Pimlico that cost £10.00 per person per night, which was very cheap, even for those days. The price even included for "breakfast in your room". Of course, by and large you get what you pay for, and we got two small rooms to be shared by 7 of us. The beds were so close together that when I had a "call of nature" in the night, I tripped over the cord of Colin Tigg's Sleep Apnoea machine and wrenched his mask off his face. It is a good job he woke up because without the device, he was prone just to stop breathing when asleep and never wake up. Furthermore, the "breakfast in bed" proved to be a boiled egg and a piece of toast left outside the bedroom door.

On the Saturday night we went out and returned to the B&B "having drink taken" as my Irish colleagues would say. Paul Barnes was supposed to be sleeping in the room with Brian, Gordon and possibly Graham Williams, but for some reason, Brian took exception to him and locked him out. After a lot of banging on doors, Paul came into out room, climbed out of our window onto a narrow ledge and inched his way along to the window of his own room. When he tried to climb in, he prevailed over Brian who was trying to push him off the ledge and got into his room upon which, we all had a good laugh and went to bed. Next morning, when I went out to get a paper, I looked up at the ledge and gasped. The ledge that Paul had used was no more than 12" wide and it was at least 30 feet in the air – and Brian had tried to push him off it! We had a serious team meeting after that about the dangers of drinking too much and pissing about afterwards.

London was to become a regular dive location for us for many years as there were 10 railway bridges across the Thames in the Greater London area alone, plus many other bridges over smaller watercourses. It was essential that we found some good, regular accommodation and one year, we hit upon the perfect place. It was an excellent B&B quite close to Surbiton Station and within walking distance of a couple of good pubs. It was run by a lovely gay man called John who became a great friend of ours despite the fact that he was affectionately referred by all the team as "The woolly-woofter". His establishment was impeccably clean, his breakfasts were legendary, and he would always greet us with tea and biscuits when we got in from diving. Despite our rough ways, he was glad to have us because he had several regular guests who stayed Monday to Thursday nights while working in London and then went home at the weekend, so if we were not there, he would have several empty rooms which he let to us at a very fair price.

One particular weekend (and not for the first time), Paul was getting the brunt of the "Mickey-taking". In the pub, Brian had convinced him that he was getting inferior beer to the rest of us despite that fact that we all paid the same money into the kitty and there had been a row. Later on, it was suggested to Paul that John was being particularly attentive to him and that perhaps he fancied him. Nothing could be further from the truth but Paul "bit" and started to avoid John when we got back to the B&B. Of course, we couldn't leave it there and John's help was enlisted in a major "wind-up". He made a point of touching Paul on the shoulder, admiring his appearance and giving him little smiles whenever he saw him which we, of course, mentioned to Paul who became more and more agitated before eventually blurting out. "You bastards, I am not fucking gay!" (Fortunately, not within John's hearing.)

Although we managed with our own inflatable boat for the Thames bridges upstream of Chelsea, for the big bridges in central London, we hired a workboat with a skipper who had the necessary qualifications and experience for such work. Generally, the jobs went well and at first, we would simply dive each pier using SCUBA gear and carry out the inspection whenever it suited us. While the tide was flowing, we would examine the abutments and side piers, but the central piers of each bridge could only be examined at high or low tide for a period of about 20 minutes of slack water. The procedure was to jump into the water whilst the tide was still running, hang onto the pier until the water began to slacken, and then start the examination and continue until the flow of the tide going in the opposite direction forced the diver to surface. He would then grab a lifebuoy that was trailed on a rope from the boat and be hauled in. Unfortunately, on one occasion a member of the public dialled 999 and reported a man hanging on to a lifebuoy in the middle of the river and the police boat was sent to find out what was going on. As well as the inspections, there were also bed surveys to undertake so the largest bridges could take all weekend to complete.

As time went on the Port of London requirements got more onerous. We had to stop using SCUBA gear and change to Surface demand equipment (of which more in the next chapter) and then we had to actually close spans to do the examination, which was difficult and time-consuming because we had to find ways to get on to the bridges at track level to hang a bale of straw over the river, a traditional sign that the span was closed. The PLA would also publish a speed restriction while we were working which was adhered to by commercial craft but totally ignored by speedboats and pleasure cruisers. Despite all this the jobs got done and I absolutely loved the challenge. The only incident that befell me was on my very last dive on Hungerford Bridge but that belonged in another era and after privatisation.

One of the Thames bridges furthest upstream that we examined was at Kingston-upon-Thames and it was generally an easy dive since it was in the very upper limits of the tidal river, being just downstream of Teddington Lock. However, on two occasions at this bridge we had a brush with the law. The first one was my own stupid fault. Two of us in a car had got to the bridge before the rest of the team in the big van and I was getting annoyed because after half an hour, there was no sign of them. I had already changed into my wet suit when I decided to set off up the High Street to see if they were getting close and to direct them to the bridge. As I walked along Kingsgate towards Richmond Road, I was aware of two men walking towards me with a purposeful stride. They introduced themselves as plain clothes policemen and asked me what I was doing in a shopping centre dressed in a wet suit and sporting a lethal looking knife in a scabbard strapped to my ankle looking all the world (as one of them put it) as an extra out of a James Bond film. I was able to convince them that I was not a terrorist threat, but I got a severe warning that carrying a knife like that in public was an offence.

The second incident also involved the police and this time they arrived by boat. We had the misfortune to arrive at Kingston to examine the bridge just after there had been severe flooding in the upper reaches of the Thames and when we got to site, it looked almost impossible for us to dive; however, having travelled all the way down from the Midlands, we were determined to give it a go. Since the bridge was located well upstream of the normal patrol areas of the Port of London Authority, we had not told them of our diving operations and expected to be in and out of the water before anyone noticed us, especially at a weekend, so we decided to give it a try. We launched the boat and started the motor but at that time, we only had a little "Seagull" engine that had very little power so once out into mid-stream, it simply went backwards, and all the crew could do was to make for the slower water alongside the bank.

I launched myself into the river and managed to hang onto the first pier. By using the protection of the pier itself, I was able to reach the riverbed and from there work my way along the bed, clinging to and rocks and debris to prevent being swept away. I have to say that it wasn't a particularly professional looking job, but we were gradually getting on with it when up came a police launch with blue lights flashing and the booming voice of a large police sergeant saying, "What the hell do you think you are doing!" Unfortunately, just then, the engine on our own boat cut out and it was heading downstream fast, so the police launch had to rescue it and tow it back to the bridge. That interlude did, at least give me the chance to make it to the bank and climb out of the river.

Eventually, the police boat was tied up and the sergeant came to the big van and demanded to know who was in charge. The first person he spoke to was Gordon whose only response was to say to him, "I am making the sandwiches for the divers, and I can't stop for you or anyone else otherwise I will be in trouble from the gaffer," pointing at me. I was a bit upset by then, so I asked the policeman what he thought he was doing by stopping us from working but he pointed out that the river was in flood, and it wasn't safe for us to work and since he was quite correct, it was difficult for me to argue the point. Apparently, while we were setting up, some busybody of a Sea Cadet Captain had reported that we were in danger although he hadn't had the guts to come down and talk to us. Once the police knew that we weren't up to no good, they rather lost interest in the proceedings and left. In truth, it wasn't possible to complete the inspection properly and we shouldn't have tried so we had to abandon that bridge for the day and pull in the dive at a later date.

Of course, these trips away gave plenty of scope for us to listen to "divers' tales" some of which were repeated regularly once the beer flowed. We also got to hear a little of the time the Brian Garner and Gordon Issott spent in the army. Brian was the teller of the tales and there was clearly some embellishment, but I think, basically, they were true. Much of Brian's service was spent in North Africa around the time of the Suez crisis although, as far as I know, he was not involved in that campaign, in fact, I am not aware that he did any actual fighting, though, that may be doing him a disservice. The only story that he told us concerned Brian and a certain Captain Jebb and the company mess. Brian had managed to wangle himself a job of helping to run the mess, under the orders of the said "Captain Jebb". The two of them were involved (with others) in ways of getting money for themselves from mess funds and were both onto a good thing until one day, the captain strode into the mess and said, "Garner, someone has told the truth and now we are all in the shit." Not long after that the Military Police were involved and a court martial was held. Brian was nearing the end of his term of National Service, but he was heading for that to be extended for about 2 years – to be served in Colchester Military prison – until it was found out that he had never passed his third-class education (or as Brian put it – Educmication). That meant, according to Brian, that he should never have been allowed to handle money or deal with accounts and therefore he could not be convicted. He always was lucky!

Gordon was not a great talker, but he had told Brian of some of his experiences in Malaya and he was content for Brian to relay them to us. Apparently, he was stationed in the jungle something like 200 miles "up country" from base. There were no facilities there and when the need for female company got overwhelming, the only thing that Gordon and others could do was to complain of toothache. They then got a ride back to base that took about 10 hours for a visit to the dentist and because of the distance at least one overnight stay. The army dentist, of course, probably knew what they were up to and insisted on removing at least one perfectly healthy tooth on each visit, hence, said Gordon, the reason for him missing several teeth. As much as Brian could be said to be lucky, Gordon was exactly the opposite and bad luck seemed to dog him.

Before joining the army, he had been living with his parents in Boston but when he returned after his 2-year stint, he found strangers living in the family home. His parents had upped sticks and moved to Derby without bothering to tell him. I understand that it was with some difficulty the Gordon managed to find their new address. So, you could say he was unlucky, you could also say he was his own worst enemy and, of course, we didn't help matters. Gordon had burst his ear drums diving before I joined the team so he was taken off diving duties and his job as an attendant, was to carry out any non-diving tasks, but this did include donning a dry suit and working in shallow water. On one Sunday morning we arrived at Bridge 27/27A over the River Trent and I said, "Gordon, swim over to the other bank with a line." Gordon said, "I haven't brought a suit, Duck," so I made him swim across in his underpants, to much of the team's amusement and a couple of early morning ramblers. He did it without complaining and whatever his faults might be he would always be there when he was needed.

When we carried out bed surveys on large rivers, we used a small sonar device that clipped onto the stern of our inflatable boat. Unfortunately, it was prone to errors when used at shallow depth. One day we were examining Kew Rail Bridge and having finished the diving, we were using the boat to take bed soundings whilst a group of lunchtime drinkers were watching us from a riverside pub garden. As we approached the bank, the sonar read 300mm of water and, fearful of damaging the prop and completely forgetting that the sonar did not read correctly in less than 1.0m of water, I stopped the motor, and I told Gordon to jump into the river with the securing rope. Gordon had on his normal trousers and jacket but was wearing wellington boots. When he jumped in, the water came up to his chest, completely soaking his clothes. On this occasion, it was not just the rest of the team laughing, it was all the watching drinkers.

While we are talking about the sonar, it was completely essential to have it where the water was deeper than about 5m. Up to that depth, we could generally manage with sounding rods, and it was Gordon's job to make sure it was taken. On a couple of occasions, he forgot so I got him to make a wooded box for it on which we painted "to be taken every time". On one fateful weekend we started work on a Thames Bridge only to find that the "to be taken every time" box had been left behind in the stores in Sandiacre. I was furious but all I could get out of Gordon was "sorry duck" so I said, "Right, you will have to go back by train for it." We sent Gordon to St Pancras to catch a train to Long Eaton and in the meantime, Colin Tigg, who had stayed at home that weekend, picked up the box (with the sonar gear inside) and met Gordon off the train, gave him the box and Gordon set off back to London. All was well but with one last throw of the dice, Gordon decided the box was too heavy to carry on the tube and the train back to

Surbiton where we were staying so he left it at the left luggage office in St. Pancras and put the sonar in his bag. That meant on the Sunday evening on our way home, we had to go through Central London to pick up the box rather than use the M25. I reasoned that there would be little difference in time but unfortunately, we got stuck for over an hour in heavy traffic while all the time Nigel, who had a dinner out arranged with his wife, was swearing, and cursing at the "so-and-so cockneys" blocking the roads. Gordon did a lot of the driving, particularly on the way back from jobs when the rest of us would sleep but his sense of direction was poor.

On one occasion on the way back from Holyhead, it was getting quite late and dark and all the team, except Gordon, who was driving, were asleep. Eventually, Alan Lodge woke up and looked at his watch. "Gordon, we should soon be home now, shouldn't we," he said. "Yes," said Gordon, who unfortunately had no idea where he was and had been driving blind for several hours, not wishing to wake anyone up to ask for help. It turned out we were near Wolverhampton having been going in the wrong direction for at least an hour. After that we tried to get a navigator to stay awake with Gordon but on one more occasion, we all fell asleep. We had been working somewhere in West Yorkshire but on the way back, Gordon had missed the turn that would take us south down the M1. Instead, he carried on east and when we awoke and discovered the error, we were in sight of the Humber Bridge. Of course, it looks funny now, but it was the cause of friction because it meant getting back late on Sunday evening and we all knew we could not be late in for our normal Monday morning duties, otherwise we might lose the "privilege" of being allowed to go diving at the weekend.

When we were in London, we would normally have a big van with all the gear and a mini-bus or something similar for the main part of the team. Gordon would be told to drive the big van and follow the minibus with the rest of us in. We knew that if Gordon lost sight of us in London, he would be lost so we would delight in approaching traffic lights as they changed to yellow, then accelerating across. Gordon would tenaciously follow us and if we timed it right, he would be crossing on red, but NOTHING would stop him. Gordon's wife Jill was a bit of a Tartar and would think nothing of shouting at Gordon in front of us when we called round to collect him for work but in her defence, she must have been frustrated with Gordon's failings.

On the final Sunday before Christmas, one year, we did a short job in the morning then went to the pub for a few drinks before dropping Gordon off at his house. He asked us in, and Brian spotted a plate of mince pies on the table. Gordon offered us one and stood by while we scoffed the lot. When Jill came in to see the empty plate, she went berserk. Apparently, she had just baked them, and they were for the Christmas family tea! We shot out the door and left it to Gordon, though, after Christmas, when we went to pick him up, I did say to Jill how nice the pies were and did she have any more! If looks could kill!

During this period, we travelled all over the Southern Region of British Rail and by and large, it was a very pleasant experience. We worked to our own rules using SCUBA equipment and regarded ourselves as a very good inspection team, capable of doing any job put to us and indeed we did get some difficult jobs. Whilst I have been quite light-hearted about the work, when we found a bridge that was not in good order, we would spend many hours examining until we were absolutely sure that we could produce a full and accurate report showing all the defects. An example of this was Glynde Reach Bridge on the Newhaven Branch. It was a wooden structure carrying the line over a small tidal river, but we spent hours there probing every piece of rotten timber in the trestles until we were certain we had all the detail we needed. An additional problem at the bridge was that there was no road access, so all our equipment had to be carried for about a mile across fields. At the end of a full day's diving, sometime after 7 o'clock, we still had to carry all our gear back to the van across the fields and load it up. On that day, we got changed in the van, headed for the nearest pub then got Gordon to drive us back to the B&B via a Chinese take-away! Other jobs were less onerous like the River Avon at Christchurch where the river was beautifully clear and full of large Trout. Fortunately, on that day, we had advised the local Bailiff what we were doing but he still watched us like a hawk to see that we didn't make off with any of his fish.

In 1984, we were on a weekend diving expedition to the south coast and as usual, we were staying in Lymington. Coincidentally, that weekend, Bournemouth AFC were playing Manchester United in the FA cup and my nephew, Ian Thompson, was due to play in the Bournemouth team. I had 2 tickets for the match so with the co-operation of the other divers, we started work very early that day and got most of the work done by lunchtime so that Brian and I could go to the match while the others finished off the work. All went well and we arrived at the ground with about 15 minutes to spare and took up our places in the Bournemouth supporters standing area behind the goal. I was a little concerned that we might be spotted at the match by the TV cameras and I didn't want to explain to the boss why I was at the match when I was being paid to be at work, so I said to Brian, "Let's make sure we keep a low profile." That match has gone down in history, Bournemouth, from the old third division of English Football beat Manchester United 2-0 and Ian scored the second goal. Needless to say, "keeping a low profile" was the last thing on my mind and when Manchester fans invaded the pitch towards the end of the game in an abortive attempt to get the match abandoned, I was seen shouting at a policeman next to me to "get them off the fucking pitch". After that high point, the diving work began to get a bit harder due to new Health and Safety legislation, but it always remained a fun job.

Water tower, Doncaster railway works

Trent Weir

Cannon Street Bridge – River Thames

Bournemouth v Man U 1984 - Ian celebrates scoring a goal

Chapter 21

The Diving gets Harder, and we move into the

Private Sector

In the 1980s, several factors combined to make diving inspections more difficult. Firstly, came the application of the 1974 Health and Safety at Work Act to diving operations in 1981. The Health and Safety at Work legislation set out to ensure that employers and employees carried out risk assessments and produced documented work systems to ensure the safety of their operations (as far as reasonably practicable) and it provided guidelines as to how that might be achieved but did not normally specify exactly how the work should be carried out. In the case of diving operations, however, it made it clear that divers should have voice communication with the surface and should have an inexhaustible supply of air in case they became trapped underwater. Except in shallow water, therefore, the only way to comply with the act was to employ "surface demand" equipment (SD). To some extent, this was a step back in time to the days of the old "standard" divers but with the use of modern, but still quite cumbersome, equipment. The Nottingham team, in the end, was the only British Railways team to attempt to comply with the new requirements and only then after considerable protestations that they added little to safety and in some circumstances could increase the risk to examining divers. In the end we lost all the arguments and so we had to comply.

Firstly, we set limits on the depth of water that divers could work in without SD equipment and these were calculated on the basis that the diver would have contact with the surface and a limitless supply of air if his head was above water when he was standing on the river bed. By this simple rule, we were able to comply with the act on perhaps 80% of our jobs without resorting to surface demand. Most SD equipment was designed for use on work boats and was of little use to us since our access to site would normally be by road with the diving supervisor working from the bank. We therefore had designed for us a portable control panel that could be carried by two people, and we purchased a telephone system that ran from a small control box that could be sited next to the diving panel. We took with us to site a portable compressor so that the air cylinders could be refilled on site but the actual air being breathed came from a bank of air cylinders connected to the panel. We initially used Comex-Pro helmets for the divers but later changed to Kirby-Morgan hard helmets. The diver's helmet received telephone communication and air down an umbilical line, which also had an open flexible "pnuemo" tube by which the supervisor could get a depth reading. Two divers (main diver and standby) could be operated from one panel. For inspection diving, the main drawbacks were firstly the weight of equipment to be carried to site, and secondly, that once the helmet was in place the diver had to continue to breathe the air from the cylinders until the end of the job whereas with SCUBA, he could spend periods on the surface breathing normally. However, by far our main objection was that the umbilical could get trapped on obstructions as we swam around the structures being examined and in flowing water this was quite likely. Some years later our fears were shown to be justified in the River Thames when I had what could have been a fatal incident.

To set the scene, Hungerford Rail Bridge carries the rail tracks (up to 6 at the station end) into Charing Cross Station. In 2002 new public walkways were added to each side, supported on the foundations of the original bridge. Part of the design for the new footbridges was for precast concrete beams to be installed at riverbed level connecting upstream and downstream footbridges and during the course of the lowering of one of the beams into position, it was dropped onto a cast iron cylinder of the original bridge causing it to be badly fractured. The main contractor for the works had to arrange for the repair of the cast iron by welding and this work was inspected by the Bridgeway diving team, led by myself on behalf of Railtrack. The Bridgeway team also carried out weekly inspections of the damaged area until the repairs had been fully completed. Repairing Cast Iron is difficult at the best of times and even more difficult when the work is having to be carried out by divers, so when the routine inspection of the structure had to be carried out about 2 years later, I was determined to be there myself even though I was approaching 60 at the time and heavily involved in the running of the company so my diving career was coming to an end. We followed the usual procedure with Paul being the dive supervisor, myself as the main diver and another of the dive team (Gary Turnbull) fully equipped and on standby. As soon as the current slackened, I was put into the water and descended to the bed. Communication systems and air lines were working well, and I was able to talk to Paul who was operating the control panel. I found that the bed of the river was a mass of construction debris that had never been cleared up by the contractors working on the footbridges and as well as loose scaffold tubes and reinforcing bars, there were numerous reinforcing bars protruding from the concrete works. My early diving had been carried out on SCUBA and this inspection would have been easy using that equipment but working on SD meant that I had to drag the heavy umbilical line with me as I inspected all parts of the structure and when I came to the area of the cast iron repair, I had to squeeze under a large concrete beam to be able to confirm that the repair work was still in good order. At that point I was told by Paul to surface because the tide had turned, and I started to retrace my route back through the maze of debris so that I could return to the boat. At the same time, the attendant on the boat was hauling in the umbilical but at some point, the line caught around a protruding bar and with both me and the attendant pulling from each end with some force, the airline was squeezed, and my air supply was cut off. Fortunately, at this point, training stepped in, and I simply opened the valve of the emergency air cylinder so that I had air to breath. However, I was not yet safe because that cylinder only contained enough air for 10 minutes, so I had to free the line in that time. In effect, to put it dramatically, this meant that I was 10 minutes from death. Paul was well aware that I had a problem as soon as it happened because his control panel showed that I was no longer receiving air from above and he had told the standby diver to be ready to assist me, but the tide was now beginning to run quite fast and the chances of him being able to help were limited, so really it was down to me. I told Paul to get the attendants to give me some slack on the umbilical and I followed it back for about 10m where I found the point where it was trapped and released it, restoring my surface supply with 5 minutes to spare. Being careful that it did not get trapped again, I then made my way to the surface and the attendants were ready to haul me back to the boat, but by this time the river was running very fast, and I could do little to help them. I simply trailed out downstream from the boat at the end of my umbilical and rested while I left them to do the hard work. Eventually they were able to haul me back and lift me on to the boat, they then removed my helmet, and I could breathe fresh air again. Thinking about it later, I was very glad that I had not encountered such a problem earlier in my career when I lacked experience because I am sure that the outcome might not have been so good.

The Hungerford Bridge incident happened after we had left British Rail and become Bridgeway Consulting, but before that, there were many jobs carried out on Surface Demand. SD was particularly useful for the inspection of long harbour walls because there was always access to the top of the wall for the "panel" to be set up and the diver could work for many hours, if necessary, without returning to the surface. Lymington harbour was a case in point and one sunny morning, we set up our gear on the harbour and started the examination. I was diving and Graham was working the panel, and I was in a particularly good mood. When things are going well on a dive, there is a particular satisfaction and even a sense of euphoria to moving weightlessly around at the bottom of the sea and I often used to sing a little to myself while I worked. I was reporting back the condition of the structure to Graham as I worked and sang but what I didn't realise was that he had the communications on loudspeaker and that a sizable crowd of holiday makers, waiting for the Isle of Wight Ferry to arrive, had gathered to watch (and listen) to the proceedings. My singing voice is appalling at the best of times so there was some hilarity amongst the watchers coupled with the interest in the diving.

We had put the panel on a trolley so that it could follow me along the harbour, and the standby diver was delegated to move it along as I worked. Unfortunately, the standby diver was the member of the team most likely to get distracted and he was off "my-chappying" to the surrounding crowd, busily explaining what his "chap" (me) was doing; so, although he was quite close to the panel and would, of course, have been immediately available in an emergency, he was not assisting in the pushing of the trolley. Because of this, I soon reached the end of the umbilical (which was 50m long) and could move no further so I naturally I spoke to Graham and my message was broadcast to all the onlookers. It went something like this: "Tell that lazy sod to stop my-chappying and push the fucking trolley, I want to get this job finished and get in the pub!" Red faces all round "topside". Fortunately for me, the ferry arrived soon after that and the onlookers moved on before I surfaced but I was ashamed to use such language in front of the small children that were excited by watching a diver at work. That wasn't the only time that a bit of "my-chappying" caused a problem and one of them was more serious and resulted in the two of us having a few words.

We were sent to Norwich in 1987 to carry out some checks on the work of contractors who were re-building the swing bridge. This bridge is unique in the UK for being the only swing bridge on an electrified railway, so it was a really interesting job to be involved with. The 25kv overhead power lines for the electric trains could not be continuous over the river when the bridge was opened for boats so a high voltage cable connecting the wires on each side had to be buried in the bed of the river. We were sent to inspect the underwater trench that had been dredged out to for the cable to ensure that it was deep enough before the cable was installed, and the trench was filled in. It was a good job that they did send us because the trench in the middle of the river had been dug deep enough originally but river currents had partially re-filled it before the cable could be installed. The contractor on the job accepted our recommendations for further dredging immediately before dropping in the cable and that was that.

I was the diver carrying out the inspection and my assistant, avoiding getting wet as usual, was on the panel while Gordon was acting as attendant. I was, of course connected to the panel by the umbilical so as I progressed across the river, this would drag behind me all the way to the far bank. The last thing that I said before I entered the water was, "Make sure you keep a good lookout for boats because if I can't get back across the river, they will have to be stopped." All went well at first but then the contractor's agent turned up and "my-chappying" soon started while I was completely ignored. No proper lookout for boats was kept. No one saw the approaching boat until it was on top of us. I knew that it was on its way because I could hear its engine as it approached, but I was waiting for it to stop before I surfaced and moved out of the way. Not surprisingly, since neither no one had given it any signals, the boat proceeded straight through the site, and I was expecting a sudden tug on my helmet as the umbilical was snagged by the boat and I finished up being dragged along to Yarmouth or wherever it was going. Fortunately, that didn't happen, and the boat passed safely over the umbilical, but it was more by luck than judgement and I was furious.

Sometimes we got special jobs to do for which the SD equipment was ideal and one of those was to inspect a stone-lined well of indeterminate depth at Belper in Derbyshire. The well was listed as railway infrastructure, and it was thought that it might have been something to do with the water supply for steam engines at some time. We had put off the job for a while until we felt fully competent in the use of SD in confined spaces so this Sunday morning, we carefully packed all the gear into the van at Sandiacre and I particularly remember saying to my colleagues, "Now we have everything we need but we don't need to take the SCUBA gear because we will be working on SD." When we got to site, we set up the panel, the umbilicals and the telephone lines and were just ready to send down the diver when someone said, "Where are the air cylinders?" I replied, "We don't need those, we are on SD..." then my voice trailed away as I realised that the air was needed for SD just as much as it was for SCUBA! My face was very red and there was nothing for it to send the van back for the cylinders because although we had the compressor on site, it wasn't safe to run the dive without a reserve air supply. To cap it all, when we finally did get started, years of debris falling (or being thrown) into the well had partially filled it and there was very little depth of water at the bottom.

Another job that we got from time to time was checking on the work of contractors carrying out underwater repairs and what we found on one occasion made its way into RAIL NEWS, the monthly magazine for railway staff. On areas where we did not operate, the underwater examination of structures was contracted out to diving companies, most of which were fine, but diving is the sort of work that can attract mavericks and opportunists and it is also a cut-throat business. One of the engineers on another Division contacted me one day to say that he was suspicious of the underwater pointing work that had been done to one of their canal bridges because he felt that the divers had not been on site long enough to do a good job. Apparently, a small diving company had initially been asked to inspect the bridge, along with others, and had reported that there were deep open joints in the brickwork from water level down to bed level. The British Rail engineer visited the site and could see the open joints at water level and so the report was accepted and the same company, under the terms of the contract, were given an order to carry out the work. To the surprise of the BR engineer, by the time he went back to site, it had been cleared and the contractor assured him that all the work had been done, indicating the new mortar in the joints just above water level. We set up to dive the site and at first, we thought that everything was in order. The water level joints had been pointed up, somewhat roughly but soundly and there were no open joints in the brickwork below water level. Bearing in mind that the water was very murky so it was difficult to see the joints and we were working mainly by feel, we were almost hoodwinked ourselves but we carried out a very careful inspection using our underwater lamps and we discovered that the new pointing could be seen to finish at just below water level and although the joints were sound below that level, the mortar in them was certainly not new. What had happened was that the freezing and thawing of the joints just at water level coupled with the wash of passing boats had, indeed caused the mortar to fail and leave deep open joints but below water level the brickwork had not been damaged in the same way and was as sound as the day that it was built. We had a situation where not only had the contractor not carried out the work that he had claimed, but it never wanted doing in the first place! We were very pleased to save some money for the company and to get ourselves some publicity in Rail News.

When I moved to Peterborough in 1989, I continued to maintain my diving competence and find time to go on weekend trips to the Southern Region, but I handed over the running of the team to others, including Nigel Stockdale, who eventually took charge in the lead-up to privatisation. I have previously explained the process of arranging accommodation but around about this time, a new system was introduced. A member of the clerical staff decided that letting us book our own accommodation could be open to abuse and so he insisted that from then on, the accommodation would be booked for us by the clerical office. We were disappointed in this because we would no longer be able to book some of our favourite accommodation, but as it turned out, this decision cost BR a rather large amount of money because there was an agreement with a major accommodation agency and all hotels were booked through them. Unfortunately, the only people that they normally catered for were senior managers. Very few other staff ever stayed away, or if they did, they booked their own hotels and claimed the allowance, just as we had been doing. Consequently, we were sent to some very expensive hotels who were not necessarily pleased to see us turn up in our filthy clothes after a day's diving. To be honest, we were not particularly comfortable either at some of them, but we did take advantage on occasion.

One of the hotels that we stayed at several times was the Balmer Lawns Hotel in the New Forest. The first time that we stayed there, we ran up a very large wine bill at the Saturday evening meal and a week or so later, when BR received the bill, Nigel was given a severe reprimand, not only on grounds of cost but also on grounds of safety since it was judged that consuming so much alcohol must have affected our ability to dive the next day. In London, we went to a very nice Hotel in Knightsbridge. When we arrived in our diving van and asked the liveried doorman where we could park, he looked at us as if we were from another planet but eventually, he found a space for us round the back of the hotel and an entrance that we could use so that we didn't need to be seen by the regular guests. We didn't mind because it meant we could get into the hotel wearing our wet suits if we wanted without being thrown out.

It was at that particular hotel that we were "allowed" by a particularly obnoxious barman to spend much more money on drinks than we could ever have really afforded on our wages. After dinner in the restaurant, we repaired to the bar and Gordon offered to get the first round of drinks. Nigel considered himself a connoisseur of whiskey and spotted a bottle of Royal Lochnager Special vintage on the top shelf, which he specified and all the rest of us, apart from Brian, who ordered an ordinary Brandy, followed his lead. Of course, we had doubles but when Gordon went to pay for them with a £20.00 note, the shit of a barman said, "Sir, that is just enough to pay for one of your drinks." We all had to dig deep in our pockets to share the cost of the round (no credit cards in those days) and we were embarrassed in front of the other customers. I was fuming that the barman didn't quietly mention the cost of what we were ordering before serving us because it would have been obvious to him that we were just working lads. That taught us a lesson and from then on, even in the best hotels, we shunned the bar in favour of the pub round the corner.

In the run-up to privatisation, the team continued to operate and indeed recruited another member, Andy Jones, but all was not well. Without a senior manager in Derby to support them it was beginning to look as if "spanner man" would get his way at last and the team would be shut down. By then, I should have been in a position, as a Senior Engineer, to intervene but old rivalries between Eastern and London Midland Region prevailed, and I was powerless. I considered taking the team under my control at Peterborough, but it was clear that if I did, the Derby-based divers would not have been allowed to remain with the team. Nigel struggled on as best he could, and they continued to do good work but at the start of 1995 it became clear that not only was the team to be shut down, but that all the diving equipment was likely to be sold-off or scrapped. By this time Pino, Andy, and I were seriously considering setting-up our own company after privatisation and we needed to purchase all the diving equipment for our new company. If that was not possible, the cost of fitting out a diving team with new equipment would be a considerable drain on our resources and in the poisonous atmosphere that was prevailing at the time, there were plenty of people only too keen to see us fail. Action was needed so in order to secure the equipment (which included large items like a boat and motor) and it was brought over to Peterborough and put into the bridge section stores where it would be under my control. Since they remained on railway property, there was nothing wrong with what we did, and the team continued to carry out some diving but had to come to Peterborough for their gear. When privatisation happened, we were able to put in an offer for the equipment, which was accepted and all of it was transferred to Bridgeway.

Of the members of the British Rail team remaining, only Pino, Andy Jones and I were ready to commit to the new company. Nigel, Colin, and Graham Williams took jobs offered to them during the privatisation process while Paul decided to go full-time diving and Brian Garner took early retirement, although all of them eventually came to work for Bridgeway and Paul Barnes and Nigel Stockdale still do. The team started operating again as a part of Bridgeway Structural Services (later Bridgeway Consulting) and the second job that came into the company was to examine a culvert at Offord on the East Coast Main Line. My good friend from Peterborough days, Ted Geeson, had an urgent need for the exanimation and new inspection contracts had not yet been set up by Railtrack so he was allowed to give us an order for the work. I think that is the only diving job that we ever did directly for Railtrack because after that, all jobs had to come through one of the large engineering consultancies that were awarded major inspection contracts. After that job, there was some time before we did any more diving work and for a while it seemed that all the knocking on doors of local authorities, port authorities and railway offices would never produce any worthwhile volume of work. The main problem was one of cost. Local authorities in particular would almost always accept the lowest tender, even when it was patently absurd and there were plenty of small diving companies prepared to ignore Health and safety to get the work.

A case in point was a tender issued by Nottinghamshire County Council for an annual bridge inspection programme, which included Trent Bridge in Nottingham. We would dearly have liked to do the job since it would have been good local publicity for us. We submitted a good tender but when the results were known, our price had been beaten by 2 other companies. One, a well-respected diving company had submitted a price that was better by a few pounds, but the winning tender was less than half of any other. I wrote to the County Engineer, explaining that it was impossible for the contractor to do the work for the price quoted, outlining the need for boat hire and working on surface demand as well as the need to have engineer/divers in the team if the inspection was to be meaningful. Sad to say, I received no reply, not even an acknowledgement but in a way, I did have the last laugh. About 2 years later, we were working on a diving job for the British Waterways Board, and I mentioned this to the local BWB manager. *"Oh,"* he said, *"that job has never been done, to work on the Trent the diving company had to submit their proposals to us and every time they did, we rejected them as unsafe"*!

Despite all the rejections, we did win some contracts from time to time, both on and off the railway but our determination not to reduce the quality of our work or to compromise the safety of our operations meant that we never actually expanded the diving operations in the same way as the rest of the business. The railway jobs that we did were generally routine inspections of the type that we had done for many years as a BR team and that, together with our very high level of railway safety and quality certification, meant that we were well placed to tender for these jobs. However, our clients were large civil engineering consultancies who had, themselves, had to tender extremely tightly for the annual inspection contracts and so we were only ever awarded work at very low margins. On a number of occasions, we had to withstand their demands to carry out the inspections at ridiculously low prices and at least twice we lost contracts because we refused to comply, only to be awarded them some months later when our cheaper rivals failed to deliver acceptable inspections and reports.

Despite this, we were able to build an underwater inspection team and the office side was run for a while by my niece, Michelle. She was critical to the operations because the client never saw us working on site but monitored closely the quality and promptness of the reports that we sent and it was her job to look after the planning and reporting side of the work. While she was with us, she took railway safety examinations and when she eventually left, it was to work for Network Rail on the high-speed track measurement train. We did manage to get some railway inspection contracts in the areas that we had not worked before.

For a couple of years, the team spent the summer in Scotland, and I excused myself from normal duties to go up there and work with them for a few days. One of the most interesting jobs in Scotland was the Clachnaharry Swing Bridge, which carries the Far North Line over the Caledonian Canal. At one time the bearings on the bridge had deteriorated somewhat so that as a train passed onto the swinging section, there was a load bang as the wheels hit the ends of the bridge rails. A regular and rather important passenger became so used to this that he would sleep as he travelled south until he heard the bang, and he would then rouse himself and prepare to alight as the train pulled into Inverness. Eventually, the bridge was repaired, and the bang was silenced which drew an irate response from the distinguished passenger when he arrived at Inverness still asleep and had to be roused by the train guard. Another contract was to examine the piers of the famous Royal Albert Bridge over the River Tamar on the Devon/Cornwall border. This time both Pino and I fancied diving the bridge and we both "played hooky" from the job of running the company to go down to Plymouth for a couple of nights. We worked off a hired dive boat from Plymouth and had a good dive in very clear water. I let Pino lead on that job because he wanted to practise using Surface Demand Diving, but I made sure that I also got a quick dip.

Away from railway work, we did get a few local authority and port jobs, particularly where the British Waterways were involved. We were able to get ourselves onto their approved list for which the standard was very high and meant that we were only competing with other good contractors, but it was still a competitive market. However, carrying out diving exams did allow us to expand into other areas of examination including railway tunnels, confined spaces, roped access and general bridge examinations. One of the railway jobs involved examining bridges across the Manchester ship canal, which proved quite difficult but interesting. The most exciting diving contracts that we won at this time involved carrying out inspections abroad. These were mainly done through UK based civil engineering consultancies although a major contract in Ireland was negotiated directly with the Irish Highways Agency. This was by far the largest non-railway diving contract that we won while I was at Bridgeway and it involved the underwater inspection of road bridges over the length and breadth of the country.

The work took place in the middle of winter and lasted for over a month. I went over with the first party to make sure we got off on the right footing and we drove from Nottingham to Holyhead in a large diving van to catch the overnight ferry to Dun Laoghaire. That brought back fond memories of working on Holyhead Harbour although the layout of the port had changed significantly, and we boarded the ferry where at one time the freightliner ships had unloaded their containers. Once we got to Ireland, we set to work straight away but some of the bridges could not be examined at the time because of high river levels and flood water. However, we persevered with the ones that we could do and towards the end of the first week, we had travelled to the far south of the country and were staying in digs in Cork. Some of the bridges were in Cork itself but the best dive was on Youghal Bridge over the estuary of the River Blackwater. This had piers in a deep channel of clear sea water and despite the cold, I enjoyed the dive which I carried out with Bridgeway's senior diver, Gary Turnbull, while Paul Barnes took charge of the panel.

At the end of the week, I returned to the UK by air while the team continued to carry out the exams. The arrangement was that they would set aside any structures that required me to take a second look and I would do that upon my return. Paul, Gary, and 2 other divers stayed on in Ireland for the whole time but Jon Gibbons who was there to take notes and write up the reports, requested a break halfway through the contract and came home for a weekend. The intention was for him to fly from Birmingham airport to Knock on the Monday morning and we had purchased a ticket for that purpose. Unfortunately, it being mid-January, the weather had turned bad and at the last minute the flight was cancelled. The airline (BMI Baby) blamed the weather in Ireland, but we found out later that all the airports were open but as they only had 6 passengers for the flight, it was cheaper for them to cancel it. Jon rang me from the airport to say he would have to go home but I was not going to let him give up that quickly and I sent him to the Aer Lingus desk where he was able to get a flight to Dublin from where we hired a car for him to drive to the west coast. Of course, this flight cost around £200.00 instead of the original flight cost of £20.00 so while Jon was checking out Aer Lingus, I had a look at the Ryanair website but at that late stage, their prices were over £300.00. Anyway, we got Jon over to Ireland and work carried on.

My visit was scheduled towards the end of the contract and by this time, the whole of Ireland was in the grip of a "Big Freeze". I caught a Ryanair flight to Knock which took off on time, but the captain warned us that we were flying into bad weather. Knock Airport is built on high ground and at the time, snow was falling but the airport said they were confident of clearing the snow before we arrived. After circling the airport a couple of times, the captain told us that we had been authorised to land and we descended through thick cloud but as we approached the ground, the pilot saw that there was still a lot of snow on the runway and immediately applied full power to the engines and aborted the landing. As we rose up, I caught a glimpse of a house roof that we seemed to clear by a rather small margin. Anyway, we diverted to Shannon, which is lower down and close to the sea, so fairly snow free and picked up a hire car there. The afternoon drive from Shannon to Sligo was one of the most beautiful that I have ever seen. It was incredibly cold; the whole landscape was covered in snow and ice and the evening sun turned it all pink. As a postscript to the eventful flight, on the way home, a week later, we were waiting in the bar at Knock Airport for the runway to be cleared of ice and idly chatting to the lady behind the bar. I told her about our flight over. "Oh," she said, "you were on that flight. I was in my house, and it nearly took our roof off!"

The next day dawned bright and clear but bitingly cold and at the first bridge I discovered that I couldn't get into my dry suit. It had been left in the van overnight and frozen solid so there was a delay while it was de-frosted in front of the van heater. Once we started, we got some rather good diving, especially where the rivers were kept free of ice by fast flows but in some we had to break the ice before we could get in and at one bridge, near the border with Northern Island, we simply walked on the ice to do all the initial work before breaking it up with sledge hammers to get at the parts we needed to examine. In shallow bays around the coast, the edges of the sea were frozen over and the weather continued like that for most of the week. One of the last bridges that we did was a very long bridge over the River Shannon, and I was able to do some prolonged diving on SD which I thoroughly enjoyed. At around the same time, we got a couple of other jobs abroad, the most spectacular of which was to examine a bridge over the Zambezi River near the Victoria Falls in Africa. Again, the work was for a UK-based consultant, and I was itching to go, but unfortunately the requirements of running the company made it impossible for me.

Paul Barnes led the team; all the equipment was air-freighted ahead, and they set off having been filled full of jabs from the doctors and been severely wound-up by Brian. Citing his knowledge of Africa from his war-time service in the mess in Cairo they went to Africa expecting everyone that they met would be out to kill them with knives, every insect that they saw would carry malaria, beriberi or some other fatal disease and the river would be full of crocodiles just ready to eat them. It is safe to say that they were not in a confident move when they set off but from the stories that they told when they returned, it was clear that they had a great time. The other exotic location that we had a contract for was to examine the two "subducting" or "sinking" bridges that cross the Corinth Canal in Greece and this time I was determined to be a part of the team. The Corinth is cut through the rock of the Isthmus of Corinth and connects the Ionian Sea with the Aegean Sea allowing craft to pass between the two without a long journey around the Peloponnese peninsula. The canal is straight with high rock walls on each side. It was opened in 1893 and was suitable for the ships of the time but most of today's commercial shipping is too large for the canal and the main usage is by pleasure craft and cruise ships. When a cruise ship or freighter passes through the canal with only inches to spare on each side, it is a spectacular event but best not to be diving at the time. Fortunately, the passage of all vessels is under the control of the canal supervisor at Isthmia, with whom we liaised VERY closely.

The "sinking" bridges do just that, when a ship is due, the road is closed, and the bridge decks are lowered down to the bed of the canal so that the ship can pass over them. It is a simple procedure that works well, and we found very little wrong with the bridge decks themselves but the concrete piles that support the bridge on each side of the canal were a very different story. Casting concrete underwater is a tricky operation and unless everything is done properly, the interaction between the seawater and the wet concrete can prevent the concrete from setting properly and leave it a soft chalky mess. I had seen this before on the "dolphins" at Holyhead Harbour and this had also happened at Isthmia to the extent that only the reinforcing cage remained in some piles, all concrete having been washed away. We were glad to find this and other defects with the bridge because it extended the length of time that we spent diving in the warm waters around Greece. At the end of the work, we drove back to Athens and just had time for a quick visit to the Acropolis before flying home. The Greek bridges were examined in 2009 and the Irish Bridges in the winter of 2009/10 and they probably represented the Nadir of my diving career. I was, by then 63 years old and although I continued diving for a few more years after that, and I see from notes that I was still reviewing reports and occasionally joining with the diving team up until 2013, my days of professional diving were coming to an end.

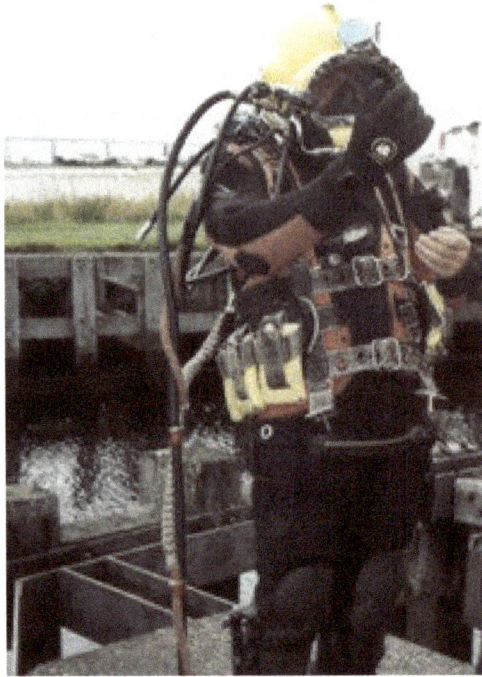

Brian in full Surface Demand at Clachnaharry Swing Bridge

Trowse Swing bridge

The diving team in 1984 illustrating different diving equipment
Front left – Brian Garner in modern dry suit with SCUBA
Front Right – Geoff Law, the division engineer, presenting Brian with a certificate for 21 years diving.
Centre left – Graham Williams in older (and colder) rubber/canvas dry suit
Centre middle – Nigel Stockdale in modern dry suit
Centre Right – Me in wet suit with SD helmet
Rear, from Left to right – Gordon Issott, Colin Tigg, Alan Lodge and Geoff Barker

Hungerford Bridge. The new walkways of the Golden Jubilee Bridges are on each side of the original railway bridge. Photo taken from the London Eye (source unknown)

Cold water work in Ireland

Youghal Bridge, Ireland. 2010

Cruise ship passing through the Corinth Canal

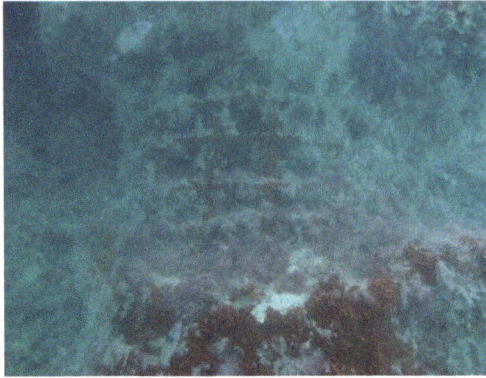

Defective concrete piles on the Isthmia Sinking bridge. Reinforcing bars but no concrete!

Isthmia Sinking Bridge

Clear water in the Corinth Canal

Chapter 22

Bridgeway Beginnings

Throughout March 1995, I continued working out my notice with British Rail Infrastructure Services to the best of my ability but to little effect as most of my workforce faced up to redundancy. It was a bleak time as I watched the organisation that I had taken such pride in not just discarded but thrown onto the proverbial scrapheap with, it seemed, undue haste and relish. I tried to save my tunnel train, only to be told that the new contractors were not interested in running trains and when I objected to unknown people taking away the drawings from my plan store, I was threatened with legal action. I think it is fair to say that the atmosphere was poisonous. It was, however, a remarkably busy time because every minute that was not spent working for BRIS was spent in the very enjoyable job of working with Andy Jones and Pino DeRosa to set up our new company. Evening after evening was spent in looking for a company name and we eventually decided that combining Bridge and Railway to make BRIDGEWAY and following that with STRUCTURAL SERVICES would be appropriate and would effectively cover what we intended to do. We then designed a logo and used a part of an arch bridge. Both choices proved successful, the name of Bridgeway is now well-known throughout the rail industry and although the logo has been professionally stylised, it remains based on a bridge arch. We obtained the services of a solicitor and an accountant to help us set up a partnership and we got good advice from the local business advisory service.

When the three of us met with our advisor, he asked us what we expected to achieve as a company. I replied that each of the three of us had particular skills that we believed would be in demand by the newly privatised rail industry and that would give us a reasonable return on our investments. At that point, Pino broke in and said, "I think we can do rather more than that, Brian," and proceeded to outline his vision of a continually expanding company that went well beyond my expectations and, indeed, well beyond what the three of us had previously discussed. I realised straight away that he was right and up to that point we had been far too conservative in our thinking. I was adamant on one point, however, that we would expand only when the time was right and without borrowing from the bank or any other financial institution. I was prepared to sink all my redundancy money and all my efforts into achieving this and since the work that we were proposing to carry out was not expected to involve major purchases of materials, I believed that it was eminently possible. I ensured that I got the support of my colleagues and the business advisor in this matter so that was how we proceeded, and it later proved to be one of the best decisions that we made. We had already decided on where our first office would be – my house!

Sometime before I had built a small ground floor extension to be a "Granny-flat" for my mother who came to live with us for a few years. By this time, however, she was in her 90s and had moved into a care home, so it was only occasionally used by us as a dining room. Since I had installed a small kitchen area with sink and hot water together with a toilet, it proved ideal for the first Bridgeway base. We had lines put in for telephone and computers by the new cable company "Diamond Cable" that had just started in the area but to save money, we used our existing house phone line for the FAX machine. This was located in our kitchen which was a little inconvenient, particularly when it rang in the middle of the night, and we heard it from our bedroom which was directly above.

The three of us were equal partners in the business and my wife carried out clerical duties. Pino's wife was also heavily involved in the early stages and her expertise in design work proved especially useful. We also recruited some IT help from a former colleague of Pino's, Mike Harris, who eventually joined the company full-time and remains with them to this day as a senior manager. It is incredible how naïve I was about IT matters, having only limited previous hands-on computer experience. In those early days, two things amazed me. We set up two workstations, each with its own PC and I was completely overwhelmed when Mike set them up to "talk" to each other and share one database. After a while helping us out, Mike moved to America in search of pastures new, but he continued to provide our IT support and I was again amazed when he solved my problems remotely from his home in the USA. The formal partnership was set up ready for the 1 April 1995, which was to be the day we would start business. We did discuss whether we should delay the start by one day so as not to for ever have "all fool's day" in our minds, but we decided that we should, instead, use it as a badge of honour to show how confident we were of success.

Before we started work, we determined on a few principles that were to guide us for many years to come. Firstly, that every contract we carried out would be given a separate job number (starting with Job 1) and secondly that every month, we would hold a progress meeting at which every "live" job would be reviewed to report on progress and ensure that the work was being correctly billed. These procedures were to ensure that, to the best of my knowledge, we never did any work that we did not charge for. The limited space available also concerned me so we bought a "scanner" and made it a rule that all documents would be held on computer unless there was some legal reason for keeping a paper copy. This proved to be an absolute boon as the company expanded and even 20 years later, other companies were amazed how well we operated without a physical "archive" of documents.

This excellent organisation was all very well but we needed work. From the very beginning, we determined to pay ourselves a basic income and we could not do that without paid work. Fortunately, we had an initial income available to us through Pino. He had left British Rail some months before and was being employed as an engineer on the Jewellery Line project in Birmingham through an employment agency. He was able to negotiate the transfer of his contract to Bridgeway thus giving us a monthly income although not enough to support the three of us for very long. Jobs for Andy and me proved a little harder to find at first and I have to say that I proved useless at canvassing for work. I was just no good at what the divers would call "my-chappying" and few, if any, of the contracts we won were through my efforts in that regard.

However, work did start to come in, mainly small jobs where former colleagues who remained in the rail industry needed the sort of help that we could provide. We got the odd small diving job and some surveys and inspection work where our railway safety qualifications proved to be a valuable commodity. Then one day I took a telephone call from a London office of the British Rail Property Board. They had been charged with drawing up plans of all stations and railway yards which would show the line of demarcation between the operational railway and land which was owned by British Rail but not required for the running of the railway. The former would become the property of RAILTRACK whilst the latter would remain in the hands of the British Rail Property Board until it could be sold-off or disposed of in some other way. They offered us the opportunity of doing this job for them over an area covering much of North and East London and out into Easy Anglia. They would provide the most up-to-date plan of each location and we would visit site; verify the accuracy of the plan and mark up the proposed demarcation lines which had to take into account not only the actual land used by the railway itself but other aspects such as cable routes and access roads which were essential for maintaining the line. There were several hundred sites to visit in a limited period but after some negotiation, we were entrusted with the job and Andy and I bravely set off armed with measuring equipment and cameras. We set ourselves challenging targets but fortunately it was springtime and light for well over 12 hours a day. Using mainly cars to get around but also occasionally trains in inner London we could cover a lot of sites in a long day, and it soon became clear that we could do more if we stayed away for a few days.

It was while Andy and I were doing this that we finished up at the most gruesome "digs" that I have ever encountered. We had been working all day and still hoped to do a few more sites before stopping for the night. I came up with the bright idea that there would be plenty of B&Bs in Southend-on-Sea eager for our business so we agreed to work on till 8pm, by which time we would have finished two more stations. We then had a quick drink before looking for somewhere to stay by which time it was nearly 9pm. After drawing a blank at several places, I saw a "Vacancy" sign and knocked on the door. We were shown a dingy room but by that time we had no choice but to accept it. The saving grace was that it was cheap but when we pulled back the bed covers to reveal the thin stained sheets we saw why. Next morning, we set off without sampling breakfast and from then on, if anyone complained about their overnight accommodation, we would simply say "Southend".

After a few months, progress was being made and although there had been no major breakthroughs in terms of large contracts, Pino was still working on the Jewellery Line, the BR Property Board contract had been completed and a steady trickle of small jobs were keeping us going. We also had our first celebratory meal which was held in a small restaurant called "Ashmores" in the next village to Bingham. We had employed the services of a consultant to set up our "Quality" systems and this had led to us gaining ISO 9001 Quality Assurance Certification. This was an important first step towards working on bigger contracts and we were proud that we had put it in place so that its principles would guide the company as we grew.

We took the consultant along for the meal together with Pino and his wife (Carolyn), myself, Ina, and Andy, who I recall was accompanied by his girlfriend of the time. It was a good evening and I think we all felt that it signalled a bright future. Unbeknown to me, our presence was being noticed in the rail industry and we got a call from Terry Everall who had previously been in the Chief Civil Engineer's office in York but who had left to form a railway division for a large civil engineering consultancy called "White

Young Green". Terry asked if we could help him out to fulfil contracts that he had won on behalf of his company. Railtrack were beginning to issue contracts for companies to design new works and major maintenance projects on the railway, but their approach was to be as "arms-length" as possible so that consultancies would also be responsible for site supervision and all railway safety arrangements, including taking possession of the line.

White Young Green (WYG) were at a disadvantage to some of their competitors in this respect because the others had taken over former British Rail departments and therefore had in-house expertise. For instance, Owen Williams and Partners had taken over the office of the BR Chief Civil Engineer for the Midland Region but WYG had no such resource to draw on. For the first job, Terry asked me if we could carry out the site supervision for a job at the former Royal Mail depot in Doncaster. The first thing that had to be done was to properly block a siding so that the work could be done without fear of a train being shunted into the site. My immediate reaction, thinking about my BR days, was to assume that Railtrack would block the line. After all, it was their infrastructure and liaison would have to be with their signallers. But I was told that from now on, Railtrack would be handing over all such duties to private contractors.

I am not sure if I realised it at the time but that one decision had presented Bridgeway with an opportunity to grow far beyond our wildest dreams, however, initially it presented us with a problem. To block a siding needs a red banner or STOP sign of an approved design and a physical barrier, normally a sleeper chained and padlocked to the rails must be constructed. We only had a couple of days to find one and we had no idea where to start to buy the necessary equipment through official channels so I took a car around some local P.-Way depots to see what I could discover, and I was lucky at Tuxford. Under a cabin, I discovered an old red banner, complete with wooded poles that had been discarded, obviously because of its condition. Rot had set into the material, so it was rather fragile, but it survived long enough to last for the length of the Doncaster job. Finding an old sleeper in the yard at Doncaster was easy enough and I purchased a length of chain and a padlock for a local hardware store. From this job, we established a profitable (but sometimes strained) partnership with WHG that was to last for the next 5 or 6 years.

The Doncaster job involved only a limited day-to-day presence on site which we managed between us but the next job for WYG was to provide full-time site supervision of a bridge reconstruction on the Hull Docks branch. That line was being upgraded at the time to handle power station coal being imported through Hull docks. Clearly, we needed to find our first employee and we approached our old diving colleague, Paul Barnes. Paul was getting limited work as a diver, and he accepted the job which included him getting to site on his own motorbike. He did a good job for us but one morning I was woken by the telephone, and it was Paul telling me that he couldn't get from home to Hull (a journey of about 100 miles) because it was snowing heavily. I am sorry to say that I was unsympathetic, and I told him that he had to go. Poor old Paul suffered badly on the journey, but he got there eventually, and his efforts enhanced our reputation as a company.

Work continued to come in from WYG and we set up purchasing arrangements with UNIPART who had taken over the stores department of BR in Doncaster, so we were able to get hold of detonators (railway fog signals used when blocking the line) and STOP boards etc. We also set up an arrangement for safety testing and certifying of our staff for railway safety duties and we began to gather a list of names of my old railway colleagues who would be available for coming to work for us on a project-by-project basis. The next significant job to come to us was both interesting and challenging. A contract had been awarded by Railtrack for extensive repairs to a very deep shaft in Beacon Hill Tunnel near Halifax together with other repairs to the tunnel lining. For months, possession of the line had to be taken every night to enable the work to proceed and for the first time WYG were entrusting us with taking a main line possession.

We recruited Mick Atter as site supervisor; Mick had been Depot Supervisor at Derby when I was at Nottingham, and I was well aware of his strengths and failings. Mick, essentially, was a selfish man who worked for his own good but would not, in the end, let you down or allow any unsafe working to take place so long as we kept an eye on him and understood that if he got a better offer, he would have no hesitation in accepting it. On the night of the first possession, I was full of trepidation. We had also recruited and trained a former railwayman to be hand signaller at the Bradford end of the tunnel to place the detonators and STOP boards and I went to Halifax to help at that end. I had no doubts that we would all do our jobs correctly, but my concern was the attitude of the signaller at Halifax box. Would he accept the credentials of 3 people he had never met who came from a company he had never heard of to take possession from him? In BR days, such transactions only took place between locally based railwaymen. The signal box was situated high above the lines, and we climbed the stairs at about midnight, quite prepared to find that the signaller had barricaded his door. As it happens, we need not have worried. Although he was somewhat brusque at first, Mick charmed him with "old railwaymen" stories and from that night onwards, he and his colleagues were as good as gold. Aside from providing the supervision, my own responsibilities to WYG were to attend progress meeting with Railtrack in York and to identify on site the extent of work to be carried out and then to ensure that the repairs were done correctly before signing off the work as complete. To get access to the shaft, abseilers set up "bosun's chair" type apparatus by which I was winched down the full depth of the shaft (something like 100m). All the time water poured down the shafts from various adits. It was a cold, wet and dirty job but I loved it, and it was most satisfactory to see the repairs taking shape. For several months, I visited the tunnel on 2 or 3 nights a week and of course, Mick was there full time.

The jobs continued to come in and some of them were a long way from Bingham, but we managed to recruit excellent ex-BR staff to help. Even in the early days, these came from as far afield as London in the South and Newcastle in the north. South of Peterborough we took charge of a major bridge reconstruction at Yaxley, and we recruited another of my ex-Peterborough staff to look after that job. On the night when the old bridge was to be blown up, my son Rob, was employed to come along and film the proceedings. The demolition was quite exciting. The contractor had hired a well-known but eccentric "expert" to carry out the job. In the lead up to the weekend he had managed to offend everyone from Railtrack's senior engineer to the local police chief and there was some threat that he would be taken off the job but common-sense prevailed and, in the end, he did an excellent (and spectacular) job in front of thousands of locals who had come to watch.

Again, I attended progress meetings with Railtrack in York and started to be a familiar figure in their office. One day, one of the managers said to me, "I don't know why we don't give Bridgeway these jobs direct, Brian, whenever we let a contract to WYG, it is always Bridgeway who do all the work." I thought it was just idle conversation, but it later transpired that they had genuinely been considering us. A second bridge reconstruction followed at Yaxley under the WYG banner and enquiries were beginning to come into us from other consultants. The most remote job that we took charge of was at Blea Moor on the Settle – Carlisle line where we looked after an underbridge reconstruction. Blea Moor signal box is located just north of the famous Ribblehead Viaduct and is one of the most remote remaining in use. There is no proper road to it and the signallers start their working day with a tramp across the moor. The contractor on the job was, in fact, the remains of a part of the BRIS organisation that was trying to make its way in the private sector so between them and Bridgeway we operated the job by co-operation in true railway fashion.

During the preparatory work I would drive from home to Ribblehead Yard and back 2 or 3 times a week, a return journey of over 300 miles! Somewhat exhausting but nevertheless rewarding. I had to be at Ribblehead Yard at 9pm where I met up with the site agent and once the last train had passed, we all piled onto an old "Wickham" trolley for the journey across the viaduct and on to the site. A "Wickham" trolley is a small, motorised rail vehicle with open sides and bench seating for around a dozen workers. The journey took around 15 minutes, and it could be beautiful on a moonlit night but was more often cold and wet. Once on site I would spend an hour or 2 inspecting the works and discussing the programme before the trolley took me back to Ribblehead Station Yard. On the weekend when the reconstruction was to take place, a Bridgeway presence was required throughout the possession which lasted from Saturday through to Monday morning. I took the first shift which lasted until around 7pm on Saturday when I was relieved by Pino. Because I was then to start back on site on Sunday morning, I had booked a room in the Station Inn at Ribblehead.

When I got there, I was served a hearty supper and then went into the bar for a couple of pints. In those days there was still such a thing as closing-up time but as this time passed it became obvious that it was not observed on a Saturday night in the dales. Approaching midnight, I told the landlord that I was off to bed because I had to work in the morning. "Aye, lad," he said, "and what time will you be wanting breakfast in the morning?"

"Six o'clock," I replied. "I must be on site soon after seven o'clock."

With that, the landlord shouted to everyone in the bar, "Drink up, lads, I am closing early tonight because this lad wants his breakfast in the morning." If looks could kill. On the second night, Pino was again providing the site supervision, but the weather changed suddenly from dry but cold to a full-blown thunderstorm. The soffit of the new bridge was only about 2m above the dry riverbed, so the contractors were using empty barrels and planks to access the bearings. During the storm, Pino suddenly spotted a wall of water tearing towards them like a Tsunami and shouted a warning to those working below the river. Fortunately, they all managed to escape but all the empty barrels were swept away and lost.

During BR days, I had never been asked to work at Christmas. In fact, Brian Davis, the Chief Civil Engineer at York had resisted all attempts by the operating department to do some major jobs at Christmas. He openly stated, "I expect my staff to work every weekend throughout the year on top of their Monday to Friday job and they deserve Christmas off." It was a different matter when Railtrack took charge, and it remains so today with many large projects undertaken over the Christmas period. My first taste of this was on another WYG job; an underbridge reconstruction at Knebworth on the East Coast Main Line towards London. It was not an especially large bridge, but it was on a key section of the line and a great deal was being made of the job. The bridge carried 4 tracks over a small access road and the plan, once the old bridge had been removed, was to carry in the new deck as one piece on the back of heavy-duty road trailers. This involved a lot of preparation work because the whole bridge had to be built up on stillages to the side of the line.

Once the weekend came and all preparatory work had been completed, the supporting trailers were placed under the bridge and the deck was jacked down onto them. Four large tractor units were attached and with much revving of engines and shouting, the bridge deck was driven into position. I have to say, having been used to bridge-jacking jobs where bridge decks were gradually moved into position along a guideway in a very controlled way, I found this to be a rather crude operation, but it was exciting, and it did work. I was doing my second shift on site when the bridge went in on Christmas morning. I was relieved by Pino and was about to head home when I took a call from Railtrack control in the West Country. The line between Dorchester and Yeovil in Dorset had been closed due to flooding and required divers to inspect the river bridges so that the line could be re-opened immediately after the Christmas break. We were contracted to be on call for just such an event and as railwaymen, we knew it was our duty to respond, but how? Fortunately, Andy Jones, having done one shift early in the Christmas period was heading to his parent's house in Sherborne in Dorset for the rest of Christmas. I got in touch with him as quickly as possible and he agreed to visit the flooded bridges straight away and to assess what we could do. He rang me back later in the day to say that he thought we would be able to inspect the bridges concerned on Boxing Day in the afternoon which would allow the line to re-open the next morning. That night, I went back to Knebworth to check on progress at the bridge and being satisfied that all was well, I left that job in the hands of our supervisor and set off to Dorset. Fortunately, the bridges just needed divers to check them out with sounding poles and the occasional dip under the water so full

271

diving equipment was not required. By Boxing night, we had examined all the bridges and declared them safe for trains. I then went back to Andy's parents' house and spent the night on the floor of their lounge. Next morning, I travelled back to home via Knebworth because I wanted to assure myself that all was well with both the new bridge and the track, and I am pleased to say that I watched several trains pass over without any problems. It was an exceptionally long Christmas away from home for most of the time but immensely satisfying in the end. I believe that both jobs enhanced the reputation of Bridgeway and led on to more work.

After 2 or 3 years of carrying out railway projects on behalf of WYG and other consultants including Owen Williams as well as a range of smaller jobs and diving work, we were in a good position, both financially and practically. But we needed to expand into new premises, and it was also decided that it would be prudent to change from being a partnership to a limited company and so BRIDGEWAY CONSULTING LIMITED was born, and the office moved from my house to OBAN HOUSE in Beeston. This was not the first premises that we looked at. Initially, I was clearly not looking big enough because I identified a house just round the corner from my own in Bingham which would have suited me to live in and it had a 2-storey annex which I thought might be suitable as company offices. Andy and Pino persuaded me off that and found some premises that were available near Trent Bridge. We missed out on those and then Pino said he had found somewhere in Beeston. I was not amused and point blank refused to go to Beeston which I referred to as a "shithole". I was also thinking about the extra hour on my working day just to get there. However, I was persuaded to look at the place and it was rather good. It was a 3-storey former dwelling that had been converted for office use and eventually we decided to invest in it. Initially, we only needed the first 2 floors, and we generated an income by renting out the top floor to a firm of landscape designers. At this time, we also carried out the change in the company structure from a partnership to a private company. On the day that Pino, Andy, and I completed the setting up of the new company, we went to the "Queens", a pub near Nottingham Station to celebrate and with us we took all the formal signed documents and our brand-new share certificates. I think it is fair to say we celebrated well and later in the evening we rolled out of the bar and, somewhat tipsily, we bid each other goodnight. Next morning saw Pino hammering on the door of the pub to retrieve the share certificates we had left behind the bar the night before!

Just about the time that we were celebrating our success, Leicester City FC had fallen on hard times and passed into administration. Gary Lineker put out an appeal for any Leicester fans who were prepared to invest money into the club to save it from extinction to come forward. Being the oldest of the three partners, I was the least cavalier with any money that came my way, unlike Andy Jones who always seemed to be able to spend his very easily. I was talking to Andy one day about the plight that Leicester City were in and said I was considering whether to help out. His words to me were, "Brian, you have the money to make a real difference and if you don't use it to help save the club, you will always regret it." He was right, of course, and I responded to the call and eventually I put £110,000.00 into the pot. Together with 42 other local businessmen, we managed to raise about 3 million pounds. It wasn't a lot, but we were unable to persuade any really wealthy people to join us, so it had to do. Unfortunately, we could only pay creditors a small percentage of what the club owed them, but we did save the club itself and I am proud to have been a part of the consortium. Eventually, we sold out to businessman Milan Mandaric and I got back just 10% of my investment. He then sold the club to the present owners, King Power. They have proved to be excellent and under their leadership, the club has gone on to win both the Premier League and the FA cup. I remain a very keen fan and season-ticket holder.

Soon after we moved into Oban House, Pino and Mike Harris announced that they intended to get fit and joined a gym in Nottingham that was housed in the old Nottingham London Road Station which had been the terminus of the LNW and GN Joint Line from Leicester (Belgrave Road), Market Harborough and Northampton and on which I had worked when it was a parcels depot. They were somewhat sceptical when I announced that I would join them because I was much older, but I persevered because I realised that during all the work of setting up a new company, I had neglected my fitness, and I was definitely overweight. The problem was that going to the gym once a week was not enough and when we did go after work, inevitably we didn't start until well after 6pm and never finished before 8 o'clock, which meant that I was eating the main meal of the day at around 9 o'clock at night and that was not good.

Then I had a revelation; the journey from home to Oban House was 13 miles which in normal traffic would take less than 25 minutes, even though it passed through the outskirts of Nottingham but in the morning and evening the journey could easily take an hour. Sitting in my car on University Boulevard one morning, waiting for the traffic to move and watching the cyclist pedalling past me on the adjacent cycle track, I realised that if I cycled to and from work, I could get all the exercise that I needed without extending my working day by more than a few minutes. I still had a bike, but it wasn't in good condition. However, I tried it out one day and found that it was possible to ride from home to office in less than an hour and all of the journey could be made on cycle tracks or minor roads. I was immediately hooked, I bought myself a decent bike and from then on rode it to work at every opportunity. This was the start of the resurgence of my cycling hobby which had laid dormant for many years and eventually led to riding "sportifs" in the Pyrenees and completing the journey from Land's End to John O'Groats. I continue to ride over 3000 miles a year and do occasional long rides of 100 miles or more.

The move to Oban House and the change to a limited company marked a watershed in the progress of the company and could definitely be said to be the end of the beginning. From then on, we began to expand rapidly, particularly in possession management, which I will explain in the next chapter.

Chapter 23

Possession Management

The term "Possession Management" was not one that I came across during my time with British Rail. In those days, when work had to be done on the line the local Permanent Way Supervisor would take "possession" of the line his own "patch" on behalf of his own track work and any other work that was required (such as bridge repairs etc.). Towards the end of BR days, the procedures became more formalised and the term "Person in charge of Possession" (PICOP) was used to do just that. Following a particularly nasty accident when a ballast train swept through a scaffold and team of painters, the idea of dividing possessions up into separate sections was devised with each worksite delineated by flashing lights and with a nominated "Engineering Supervisor" appointed. After privatisation and with many different companies carrying out work on the railway, all PICOPs and engineering supervisors had to be fully trained and certificated and had to work for an approved company. It was this world in which Bridgeway was working and when full Railtrack certification was attained, and it was a world in which it was to become a major player. In fact, I believe that it might even have been myself and my colleagues at Bridgeway who first used the term "Possession Management" to describe the whole process of planning the possession, carryout out risk assessments and producing all the necessary documentation for the job, then supplying the trained staff. It also covered all the arrangements for authorising the start of work at individual worksites and authorising the movement of ballast trains into the possession and through the possession. Finally, they had responsibility to liaise with the signaller to allow ballast trains out of the possession and to hand-back the line to the operations department once they had ensured that it was safe to do so. It was an onerous task that personally caused me many sleepless nights when I knew that we had very complex operations in progress, but it also proved to be a major source of work (and therefore income) to the company and remains so today.

Sometime after the move to Oban House we began to put the procedures in place that would allow us to become Possession Management suppliers directly to Railtrack rather than as sub-contractors working through the main contractors who were carrying out the work (such as WYG for instance). We already had ISO 9001 Quality Assurance and we now worked hard to gain IS0 14001 (Environmental assurance) and the very necessary "Contractors Safety Case" required by Railtrack for all major contractors working on rail. We also became "Investors in People", thus demonstrating our commitment to our staff. I was the director mainly involved in these initiatives, especially the "Contractors Safety Case", which involved detailed documentation and extensive auditing by Railtrack assessors. Getting the safety case also meant that we were approved for the Railway "Link-Up" scheme so that our name would be available to companies throughout the industry.

All of this took a lot of my time for the best part of a year, but I was successful in the end and not a moment too soon because Railtrack were about to offer us a massive opportunity. I have mentioned that we had been "sounded out" about working directly for Railtrack by their Eastern Region office in York and lo and behold, within days of each other in 1999, invitations to tender were issued to us to participate in the two biggest work projects that would be carried out on the region over the next 3 years. The two contracts we would tender for were to provide possession management for the projects and for a company of our size, these were massive jobs. Both invitations were from Railtrack's York office and the first to arrive was for the redevelopment of Leeds station. The scheme which involved station enlargement and rebuilding together with a complete revision of the track layout was to take over 3 years to complete with possessions being taken every night and at weekends. Furthermore, the possession would be complicated because not all lines through the station would be closed, there being a requirement to maintain the Manchester Airport to York overnight service, amongst others.

Pino, Andy, and I discussed whether we could resource the job but this time it was me that said, "Of course we can," and so we decided to bid for the work. I am not sure who the competition was for this job but since we were leading the concept of possession management, we were in pole position and were duly awarded the contract. All we had to do now was to find a good on-site manager and about 20 new safety-critical staff who would have to be fully trained and familiar with the track layout in the area. However, before we could draw breath on that job, in came the second invitation – for the Sunderland Metro project which involved electrifying the Newcastle to

Sunderland line so that the Newcastle Metro system could continue through to Sunderland and beyond by sharing the tracks with the existing Railtrack services. Again, it was a project that would take years to complete, involving not just the electrification but new stations, new signalling, and a complete change of operating procedures. This time, I had a harder job to convince Andy and Pino that we should go for the job. Newcastle and Sunderland were not areas that we knew well, and it would mean a further expansion of our management and staff resources. However, I was determined to go for it and go for it, we did. Not only that but somewhat to our surprise, we won that job as well.

The Leeds job was the first to be set up, although, if memory serves me correct, it was Sunderland that had the first actual possession. We decided to ask Mike Harris if he would manage the Leeds job. Although he had joined the company to be our IT expert, Mike had already become involved in weekend possession work and had proved himself excellent at the job. It meant a long trek for him from Nottingham to Leeds on most nights of the week as well as daytime meetings with Railtrack and contractors. Andy Jones would stand in for him on site when necessary and Mike and Andy recruited some excellent senior possession staff who could be relied upon and who would also stand in for Mike when necessary. We set up camp in Portakabin offices about a mile west of the station at Whitehall Junction. Complex possession work like this is fraught with problems and is very unforgiving. Mistakes made on the ground will, at the very least, mean a delay to the start of the night's work and at worst could result in a serious accident and loss of life. We mitigated against the possibility of the latter by having a robust system of planning, training, and the distribution of accurate information to the staff together with regular audit checks.

At Leeds we had very few safety critical incidents but I can recall one. This would have been comical had it not caused train disruption and embarrassment. Normally the arrangements for stopping trains on a particular line involves the signaller turning the necessary signals to danger then the hand signaller placing a STOP board and "Dets" on the line as a physical indication to the driver of any train that approaches the site. If the driver doesn't see the STOP board for any reason, then the "Dets" will explode under the wheels of the train. However, there is another way of doing it in certain circumstances that does not involve a physical warning to be placed on the track. In this case, having agreed with the signalman that the necessary signals have been turned to danger, a qualified Signal Technician will remove a "link" in the lineside signalling cabinet that will ensure a RED aspect remains displayed on the signals concerned. For a particular job at Leeds, it was decided that this was the most appropriate way, and the operation went well on the first 2 nights. On the third night, however, at the end of the possession, the signal technician dropped the "link" and it rattled down in the cabinet, unfortunately shorting out a lot of circuits in the process. This caused trains to be stopped abruptly over the whole station area and major disruption to service. That got us reprimanded as a company because of the amount of disruption caused but fortunately, everyone accepted that it was simply a case of cold fingers! I attended a lot of meetings with Railtrack during the course of the project and I also carried out quite a few site audits, including one of our thespian hand signallers who would only work for us when he was "resting" in the theatrical sense. That particular site was at a junction close to the station and the hand signaller had to demonstrate to me how and where he would place the protection at a very congested location.

Then there was the Sunderland Metro project with office space allocated to us in Sunderland itself. It did not require quite so many staff, but it still needed a good manager and this time it would have to be someone who lived locally. Again, after a bit of searching, we found just the man in Tony Tweddle who I had met previously during BR days. He did an excellent job for us and was able to recruit some very good staff. He was assisted by my own son, Robert, who was living in the Newcastle area following his graduation from Newcastle College. Rob had been doing part-time work for us while studying and now worked regularly on both the Sunderland job and the Leeds Job. We needed a good team at Sunderland because the main contractor had little knowledge of railway working and we had to be very firm with the contractor's staff. The line passes through some rather dubious areas and Rob told that one night, while waiting at a particular signal for authority to place protection on the line, he was bombarded with bricks being thrown from the retaining wall above. He had to shelter and wait for the police to arrive before he could venture out onto the line. Again, I paid site visits and attended meetings on the project, occasionally staying with Rob overnight. The Sunderland project was eventually completed and our success there and at Leeds set us up for many more projects, including a very large presence on the West Coast Main Line upgrading project and following the use of Rob at Leeds, this time it was Andy DeRosa (Pino's brother) who took centre stage.

We were now providing Possession Management Services all over the old Eastern and Midland regions, both directly for Railtrack but also to some of their main contractors. We had 2 full-time staff organising the manning of all these jobs and preparing and issuing the detailed plans. Staffing the weekday and weeknight jobs was rarely a problem but weekends were another matter. Generally, our very able planners were left on their own to staff the jobs on Monday to Thursday but on a Friday morning, I would be seen in their office being briefed on the weekend shortages. It would not be unusual for Friday morning to start with us being around 20 staff short for the weekend and then the work would start.

Firstly, I would look at all the jobs to see if we could reduce the numbers anywhere without reducing performance and then see if any of the staff had the time to be used on more than one job. If that didn't solve the problem, there would be a lot of ringing round and arm twisting of suitably qualified office staff. Finally, if necessary, Pino, Andy and I might have to go out and help. Somehow, we always managed to sort it but sometimes it was a relief if a client rang up and cancelled a job. Of course, not everything was plain sailing and mistakes sometime happened. Usually, it was a case of someone getting lost on route to a job and a short delay but very occasionally, a hand signaller would place protection on the wrong line, resulting in trains being delayed and a reprimand for Bridgeway but, more seriously, the person would be putting themselves in danger of being run over by a train. We worked hard and long to eliminate these instances by introducing extra training and by checking and double checking all the information sent out in the "site packs" and we did manage to reduce the number of instances but there would still be the occasional problem.

One instance that I hadn't anticipated happened in the London Area on the East Coast Main Line. A PICOP named Terry Pegram worked for us in the northeast and up there, his broad Geordie accent was no hindrance to him, in fact, it was almost certainly a help but one weekend when there was not much work on in the Newcastle area, Terry agreed to travel down to Potters Bar to take a short Saturday night possession. Had the possession been at an old-fashioned signal box where Terry could have met the signaller face-to-face, I am sure there would have been no problem but the communication for this possession was with a signaller at King's Cross power box who happened to have a strong London accent. The upshot of this was that the signaller deemed it impossible for Terry to take possession because neither party could be sure that they fully understood what each was saying. This resulted in nothing more serious than the loss of a night's work and I could see both sides of the problem. After that we tried to be more careful when sending staff off their usual patch.

Another, much more serious incident occurred in the Crewe area. We were managing a possession on the line between Stoke-on-Trent and Crewe. Towards the end of the job, a ballast train was sent down to the blocking point at the junction with the main line and stopped at the protection where we had a fully trained but somewhat inexperienced hand signaller waiting. The PICOP advised the hand signaller to tell the driver to proceed to a particular signal number and then to contact the signaller for permission to go on. The driver was advised of this, the protection was removed, and the train departed. The trouble was, the signal number quoted to the driver was only a couple of hundred yards up the line and this is where that old devil ASSUMPTION played its part. Our hand signaller ASSUMED the driver knew where the signal was, and driver ASSUMED that the hand signaller would have said "the next signal" as well as quoting the signal number so he carried on straight past the signal and potentially in the path of on-coming traffic. Of course, the mistake was immediately flagged up at the power box and an emergency radio message was put out for ALL trains in the Crewe area to stop but for some reason the driver of the ballast train ignored the message and carried on for miles, looking for the signal number that he had been given and wondering why all the other trains had stopped. There was a major enquiry into the incident and some criticism of Bridgeway staff, but most was directed at the driver. It badly affected our young hand signaller, though, and he left the industry shortly after.

I have to say that the stress of having so many staff working on possessions at weekends began to affect me and I dreaded picking up the phone in the middle of Saturday night in case we had a serious (and possibly fatal) incident on our hands. It never happened but I was pleased not to have to deal with it anymore once I had retired. I think too that it might have been one of the reasons why Andy Jones retired as early as he did. Of course, it was not always so fraught, and we did a lot of good work. I know we had a better record than most other companies because we were frequently asked to spread our wings into other parts of the country and when one of our competitors was taken off their Thameslink contract in Kent, Sussex and South London, it was to Bridgeway that Railtrack turned to fill the gap.

Following the tragic derailment at Hatfield (due to a broken rail), Railtrack gave way to Network Rail, but Bridgeway continued to provide possession management services to the new company. During the latter part of the Railtrack period, we began to feel that our success had bred some resentment in the minds of our client. It manifested itself most obviously in the new contracts that we were allowed to bid for. We were providing possession staff to Railtrack in the areas that we operated but suddenly, we were advised that we could no longer tender for that work and instead, we could bid for work on the former Great Western Region. We saw no reason for this at the time, but we looked on the bright side and decided that maybe it was to enable us to spread our good practises to other parts of the country.

We responded in typical Bridgeway fashion by winning the contract that covered most of South Wales and the Western region lines from London to Penzance. For this we set up an office in Bristol. We were becoming a national company and we celebrated by organising a charity cycle ride from our Bristol office back to Beeston with all proceeds going to the "Railway Children" charity! Over the years, it was also noticeable that when some of our competitors struggled to meet the requirements of their contracts, Bridgeway were often contacted to fill in and gradually our foothold in our local areas returned. As well as working for Railtrack and then Network Rail, our Link-Up qualification meant that a wide range of companies came to us for the service. These would be contractors working on specific sites and maintenance contractors who were working regularly on the railway throughout the year. Some of these were good clients and some were spectacularly bad but both sorts offered us opportunities and we gave all of them a full planning service, not just putting "bums on seats". The extra cost of doing this did put a small dent in our profit margins and sometimes we lost work through being undercut but it was well worth it for the customer and for our own piece of mind. Sometimes it was the bad clients that made us the most money because we would have to provide extra resources to cover the client's poor performance. It was also noticeable the while some clients were very cost conscious and would always try to get us to do "more with less", others were very cavalier in this regard.

We had work every weekend for Birse, who were works maintenance contractors to Network Rail in the Midlands and whatever their possession requirements they would always request the maximum number of site staff to ensure that the possessions were taken as quickly as possible, thus giving them the maximum of site time. However, some of their smaller jobs didn't last long so the frequently finished early. We wondered why they didn't cut down on hand signallers (for instance) to save a little money if time wasn't critical. Then the penny dropped. In their contract with Network Rail, they had to source the site safety staff on a cost-plus basis. The more staff that we provided, the more money they were paid. It was certainly a good contract for Birse and Bridgeway, but not so good for the rail industry as a whole or the paying passenger!

My personal involvement in the possession work dwindled as we were able to employ more experienced staff, but I did get heavily involved for a time with our biggest contract which was for the West Coast Man Line upgrading. A massive project and one which required us to set up an office in Rugby. I would occasionally visit when I was on-call and meet up with Andy DeRosa, our manager for the project. The work was hugely complex as was the management structure which had us working alongside the Project Management Team, which was provided by an American company. The Americans were great guys and had lots of stories about working on the American Railroads, they also like football (soccer as they called it) and I was pleased to take them as guests to Leicester City from time to time. I did still go out to place protection, right up to when I retired. These were mainly when my older son, Tim, was doing some extra work at the weekend as a PICOP and I would be working for him. I really enjoyed that; it was good to be working together.

My younger son Rob also moved down to the Nottingham area from the northeast and I would occasionally work for him as well. Those occasions brought me back to the reality of the work that our possession staff did night after night. Firstly, driving to the right access point and getting there in good time. Then contacting the PICOP for instructions and carefully checking that I knew which lines that I had to place protection and when. Then sitting waiting while the last trains past through before getting the call to place the protection. We were expected to get the line blocked within 10 minutes of getting the GO and since the blocking point and access point might be quite far apart it was a rush to complete on time. Once the protection was down and I had informed the PICOP, I always double-checked my measurements to ensure that I had everything in the right place. Sometimes the hand signaller had to use a special key to turn signals to red and before doing so there was a strict sequence of wording to follow with the signaller, always remembering to use the "NATO" phonetic alphabet.

Once, near my old home in Peterborough, I had to contact the Peterborough signal box to do the job and I had a mental block. All the signals in the Peterborough area are prefaced with PB which of course had to be spoken as PAPA BRAVO. Unfortunately, I referred to it as PLUTO BRAVO and the signaller made me go through the whole message again like a naughty schoolboy. He had to do so, of course because all communications were recorded and in the event of any incident, one of the first actions was to "pull the tapes" in the signalling centre and check every word that had been said.

Possession Management was the mainstay of the company for many years, and I am pleased to say that Bridgeway continue to do this work now. I know this because I see my next-door neighbour setting off most Saturday nights in his Bridgeway gear to do a hand signalling shift, just about the time that I am going to bed.

Chapter 24

Money Matters

This will be a short chapter, after all, money and tax matters hold very little interest for most people (me included) but I mentioned earlier in the book that I was determined from the start that our business would not be beholden to the banks so we would provide the necessary funding from our own resources and that in the first place, this was my redundancy money from BR. After that, we relied on our earnings and that was all very well as long as we got paid properly by our clients but, as anyone in business will tell you, some clients seem determined to pay as little and as late as possible.

In the early days, we didn't have this problem. The large consultants that we worked for such as White Young Green, generally paid promptly but as time went on and we began to serve a more diverse customer base, we soon found that some of the big contractors in the industry were appalling. I have no hesitation in saying that the worst of these by a country mile was Jarvis, for while there were other bad payers, Jarvis had it down to a fine art. During the privatisation process, when track maintenance and renewals work was being parcelled up and handed out, most of the contracts went to established civil engineering companies such as Balfour Beatty and a very few went to management buy-outs but to our surprise, a large amount went to a company I had never heard of called Jarvis and to this day, I don't know much about them before they came into railway work. Suffice to say, they soon earned themselves a reputation amongst railwaymen for treating their staff (and particularly the ones who were transferred from BR) exceptionally badly. Before long we began to work for them in many parts of the business including inspections and possession management.

Then the problems began to arise. At progress meeting after progress meeting, jobs for Jarvis remained on the agenda long after they had been completed because we had not been fully paid and because both Pino and I were adamant that we would not walk away and we would get our money. When our finance section could not get a proper answer, then Pino and I would try to contact Jarvis finance managers but to say that they were evasive would be putting it mildly indeed. A typical scenario would be that payment would be overdue by at least 3 months and I would contact (not for the first time) their finance manager in York who would sound sympathetic, ask for more details from me about what was owing (despite having been given it many times before) and promise that at the end of the month we would get a payment.

Sometime after the end of the month, initially to great relief, a lump sum payment would arrive in our bank account, but it was always smaller than we expected. Our finance managers would try to allocate the money to individual jobs only to find that there were no details of what Jarvis was paying for. In this way, Jarvis attempted to reduce the money they paid us by obfuscation and confusion. Only because we were well organised and tenacious did we eventually get paid all that we were owed but if we had been a smaller company, we would probably have had to accept what they had paid and write off the rest. Because of all this we began to stop working for them, but this proved impractical because partnerships were being set up by Railtrack that involved Jarvis that we needed to be involved with. An example of this was the West Coast Main Line when Jarvis and others were involved in a joint organisation which awarded us many contracts including a complete survey of the whole route. Trouble was that it was Jarvis who were the paymasters and so the problems of getting paid continued.

One particular job that we were involved in was the construction of new station platforms. At the end of the work (which we were rather proud of) we fought long and had to get properly paid but Jarvis (having caught a cold themselves on the contract) simply refused to pay in full and instead offered us partial payment for a quick settlement. Since we did not need the money immediately and had no bank manager on our back demanding interest payments, we were able to hold out for the months that it took to get all that we were owed, but eventually we did, following the threat of legal action. I know of at least one company that went out of business through not being paid properly by Jarvis and several others who had to walk away with far less than they were owed just to get a settlement. All this didn't do Jarvis any good in the long run because they eventually went out of business themselves and I for one would say good riddance.

Railtrack themselves were difficult to deal with sometimes and at one stage they hit on a device called a "reverse auction" with which they sought to get the lowest possible rates for the supply of labour. They issued tender documents specifying the grade of workers that they needed from basic labourers through to skilled staff and I dutifully went through all our figures to achieve what we considered were fair rates that would cover wages, transport, national insurance, etc., and leave us with a reasonable margin for profit. Having provided supporting documents, we were accepted for "the reverse auction", which was to be conducted on-line and involved each company bidding lower and lower prices until all but the lowest dropped out. I was at an important meeting in London that day so it was agreed that Pino would submit our prices and he had all the information that I had produced showing the lowest prices that it would be prudent to use.

By the time the online auction started, my meeting had finished, and I was on the train at King's Cross Station. Pino contacted me on my mobile and we attempted to keep the line open as the auction proceeded. Of course, this was impossible at first because there are 9 tunnels within 20 minutes of leaving London, but we did our best and it soon became clear that competition was going to be fierce. The prices for basic labour got lower and approached our bottom line but by then we had entered into the excitement of the auction and when the time came when we should have withdrawn, we carried on and eventually decided to submit one more really low bid. Just then I lost contact with Pino and was left fretting for 5 minutes or so because I realised that we didn't want the contract at that price! Fortunately, we were again undercut and dropped out. Later in the auction, we did pick up some of the work, mainly in the more technical grades which we were better equipped to provide. Fortunately, Railtrack soon abandoned the procedure, so we didn't have to repeat the action, for which I was grateful.

In the very early days of the company, I was concerned that we might make mistakes with our tax returns but fortunately, we found an excellent accountant to keep us on the straight and narrow named David Howarth who served us well for many years until he felt that we had grown too big for him, and we had to use a larger accountancy firm. It was a pleasure to work with David, not least because of his love of music. He needed to do a good job because as well as employing people directly, the rail industry also used a great many self-employed people who worked for several different companies. Making sure that we got the tax right for them was very important because HMRC kept a very watchful eye. Indeed, we did get a few visits from the "TAXMAN", the first of which was quite disconcerting.

About 4 months after we had moved office from my house to Oban House, I happened to be at home when there was a knock on my front door and there stood 2 men in grey suits and long coats, demanding to be let in. They said that we had made no recent tax returns and demanded to know why, quite aggressively. It took some time to convince them that the company had submitted returns, but from our new address and eventually they left, never to be seen again. It was quite frightening at the time, and I am glad that Ina didn't have to face them on her own. The next visit was from the VAT men who spent a day with us. Fortunately, again, David had done a good job and the only mistake that they found was in our favour and we got a small rebate. The final visit that I was involved in was from HMRC and was again, very aggressive. The circumstances left a very nasty taste in my mouth. Without going into details, they accused us of illegally employing a good friend of mine without declaring it for tax. It transpired that they had received an anonymous letter with detailed allegations that could have only come from someone who knew both me and my friend and the work that Bridgeway did. Our records were able to prove conclusively that the allegations were false but despite me pleading with them not to visit my friend (he had mental health problems) they still went ahead, and it caused him a lot of upset. To this day, I don't know who the accuser was, but I believe it could only have been someone in the local church community. Not a particularly Christian attitude!

As we grew as a company, we appointed managers to look after our finances, one of which was my own son Tim and I was able to distance myself from it to concentrate more on the engineering, much to my great relief. However, I continued to attend the monthly progress meetings, so I remained in touch with the financial health of the company until the day that I retired. It was when we were looking into how to deal with my retirement that our accountants introduced us to a company that wished to buy us out. The money that they offered was mouth-watering and we were briefly tempted to sell up and live the life of luxury. However, once we had met the company's representative, it was clear that their plans for Bridgeway did not meet with our values and would almost certainly result in loss of income and probably employment for many of our loyal staff, so we rejected their offer and told our accountants that we had no interest in a sell-out. Instead, we devised a scheme whereby I sold my share of the company in return for a fair sum, payment of which was spread over a number of years. This all worked well, and it gives me great pleasure to see the company continuing to prosper under Pino's guidance.

Chapter 25

What else we did at Bridgeway

When I wasn't being involved with the running of the company or dealing with possession management matters, I did get the chance to be involved with the myriad of other types of work that Bridgeway were undertaking, particularly ones that involved Civil Engineering in some way. I had several railway engineering qualifications that were often in demand. One of those was as a tunnel engineer and I have covered this in Chapter 11 but I was also a "Bridge Strike Examiner". A bridge examination engineer competent to check and sign off reports and make recommendations – known as an STE2; a railway hand-back engineer able to hand-back the line after engineering work and to raise and lower speed restrictions (that one thanks to Alex Turner's training) as well as various other more obscure qualifications, including being able to carry out periodic re-assessment of bridge examiners.

The bridge strike examiner was a useful source of income both for myself and my company because we were paid by the visit and, although I was paid only £50.00 at first, the company got twice that sum. When a bridge is struck by a road vehicle (or on very rare occasions, a boat) it must be examined as soon as possible thereafter by a qualified "Bridge Strike Engineer" (BSE). Exactly what happens between the bridge strike being reported to Railway "control" and the examination by the BSE will depend on the bridge itself. If the bridge is inherently strong or if it is protected in some way (known as a green bridge) then the driver of the first train will be instructed to pass over the bridge at caution and thereafter, trains will resort to running at line speed. At the other end of the scale, if the bridge is of a type that could be damaged by a bridge strike, then all trains are stopped until a BSE can get to site.

We were contracted to attend bridge strikes at specific vulnerable bridges which were close to our homes and my particular group were situated on the East Coast Main Line at Grantham. These have now been fitted with protection beams but until they were installed, trains were kept at a standstill until I arrived so if it was the middle of the night, I would throw my PPE over my pyjamas and drive the 15 or so miles to the site as quickly as I could. When a vehicle hits a railway bridge it is almost always the vehicle that comes off worst and I never actually attended a bridge that was so badly damaged that I could not immediately get the trains moving. However, there were occasions when I would report damage that would have to be repaired later and I would also get details of the lorry company and driver to include in my report. These were normally taken from the driver but if he had left site, I would contact the civil police for details.

On my first "bridge bash" call-out, I approached the site with trepidation, expecting to see mayhem and carnage but in fact it was normally a low-key affair with a few paint scrapes to the bridge and possibly a badly damaged semi-trailer so the job became somewhat routine, though it was always quite exciting driving as fast as I legally could to get to the site then sorting it all out with York Control. On just one occasion, though, I picked up the phone at home and the controller at York said they had a bridge bash for me. However, straight away, he said, "Brian, I don't know what you will find when you get there. All we know is that a double-decker bus full of children has tried to get under the bridge, but we don't yet have any other details." As you can imagine, my heart sank and all the way to the site I was fearing the worst. As I approached the bridge, the road was sealed off by police, but I was let through when I explained why I was there. At the bridge, I was presented with the sight of the bus – minus roof – on one side of the bridge and the whole roof on the road under the bridge, but thankfully, no ambulances. Apparently, what had happened was a double-decker bus with about 60 small children on board had attempted to drive under the bridge, but as it approached, one of the teachers appreciated what was happening and shouted a warning for the children to crouch down and this action saved all the children from serious injury, thank goodness. Needless to say, the bridge was undamaged.

The school trip double-decker at Grantham

"Handback" engineers were also heavily in demand, particularly from contractors who were carrying out any possession work that might have disturbed the track. It would be a requirement of their contract that they would employ a "Handback" engineer to sign off the line as fit for traffic, either at "line speed" or any agreed temporary speed restriction (TSR) before possession could be handed over. Since some of these contractors were in competition with us, we felt obliged to charge a high price for the service unless there was a good commercial reason to do otherwise. The jobs were normally quite routine and only needed a brief site visit at an agreed time before the end of the job. Others were more onerous and required staying on site for the duration of the job to make sure that excavations were not disturbing the underlying stability of the track and occasionally, there were more interesting and complex jobs.

One of these was at the large bridge at Althorpe, near Scunthorpe on the main line from Doncaster to Grimsby. This line carried not only passenger traffic but also heavy freights to the steelworks and oil trains from Immingham dock. It is a large bridge that once had a lifting portion, though, this had long been "fixed" in the down position and as well as two railway lines, it also carries a road. A contractor (who had better be nameless) had renewed all the "longitudinal timbers" on the bridge. These are large timber baulks which carry the load from the trains down to the bridge structure. The baulks must be solidly fixed to the girders below and to be very accurately positioned because the rails are directly fastened to them, and any movement or inaccuracy will affect the stability of the trains passing over them.

I was contacted directly by Railtrack one morning and told that the work had been carried out by the contractor over the previous weekend but Railtrack's own handback engineer had imposed an emergency speed restriction of 20 MPH and requested remedial works be carried out. These works were due the same night but as they had no one available, would I go down and inspect the track after the work and raise the speed to the ruling line speed of 50 MPH. I should have smelt a rat! Really? Did mighty Railtrack have no one available for a couple of hours early on a Tuesday morning or did nobody fancy it.

Anyway, I got to site about 2 hours before the end of the possession and the contractors were already packing up their gear. I introduced myself and set about walking through the tracks, cross-level gauge in hand. What I saw amazed me. Not only were there significant "twist" faults (one rail higher than the other) but what improvements in level had been achieved were by inserting a motley selection of timber packs, metal plates and other odd bits of hardware under the baseplates. It was a complete mess and far from raising the speed limit, it was touch and go if I could even open the line at all. In the end I opted to lower the speed limit to 5 MPH much to the annoyance of the contractor's man on site. Of course, once I let control know the situation, all hell let loose because a 5 MPH speed restriction requires special measures to be in place to warn train drivers and until they could be implemented, every train had to be stopped and the driver warned to cross the bridge at 5MPH. I got a call from a Railtrack engineer (suitably grumpy) and he came to site to see for himself. Once he saw the job, he was equally mortified. That contractor was not offered any more "long timber" jobs. I do have a little sympathy for them because I know from my own time with BR and later with Bridgeway that they can be very difficult jobs, even for an experienced team.

Sometime later, Bridgeway acquired several P-Way gangs and one or two good carpenters, and we did a few long timber replacement jobs ourselves. We replaced a fair number on the famous Barmouth Viaduct in Wales. Pino took charge of that one and did a pretty good job. We also replaced a set of timbers on a bridge inside West Burton Power Station. These were difficult because the track was on a tight curve and special fastenings had to be used to retain the horizontal forces imposed by loaded coal wagons crossing the bridge. The one that I remember most was in Nottingham at the time that the new Nottingham Tram system was being installed. To accommodate the tram lines in the Bulwell district, the Robin Hood line from Nottingham to Mansfield had to be reduced to a single line in places and whilst this was done, it was arranged to renew the long timbers over a bridge crossing the River Lean. Our own surveyors carried out a detailed survey and we produced designs for the new timbers. The timbers themselves were of "Greenheart", a dense timber from tropical rainforests which we obtained complete with "sustainability" certification. The main difficulty with long timber jobs is always getting the old ones out. They are often fixed into position with pitch or concrete, all of which must be broken out. That was the case on this job and while I was watching our gang attacking the concrete with compressed air breakers, I was not impressed and thought I could do better so, rather like Jack Kilsby on that early job of mine in Harringay, when one of the lads took a break, I picked up his breaker and set about the concrete with gusto. I thought I was doing a pretty good job, and so I was but when I turned round, there was a full gang of about 10 men happily watching me doing their work. Just then, one of the younger lads who didn't know me piped up, "Who's that c*** on the breaker?" Needless to say, his mates quickly shut him up and when I handed back the breaker, they all got on with the job and soon got things moving.

A perk of being a well-qualified and experienced railway engineer was that I was frequently in demand to assess other engineers on their competence, and this gave me some interesting away-days because it was much easier for me to go and assess examiners and engineers in their own workplace than have to set-up a site for them to come to Nottingham. One regular trip was to Glasgow and my preferred way of getting there was to take the early flight up from East Midland Airport, start the assessment at about 9am, finish mid-afternoon, and then take the train back home, taking advantage of the restaurant car for my evening meal. I also had the job of assessing tunnel engineers based in Croydon. For that, they arranged for us to use a tunnel on a preserved railway and when we arrived, we had to wait for a steam train to pass before we could start work. I also had the job assessing bridge and tunnel examiners who worked on "closed" lines and we spent a very pleasant couple of days on the old "Somerset and Dorset" line to the south of Bath.

At one stage, Bridgeway were running about 4 track gangs under the control of Karl Crompton who not only looked after the gangs but also had an eye for picking up work from a whole range of clients. All the gangs came from the Maltby area of Yorkshire and many of them had once been miners. They were a decent lot in general but certainly a bit rough and ready. They also had the disconcerting habit of calling each other "Bert". Consequently, I never knew the real name of many of them but of course they knew mine. I had always prided myself on knowing the name of everyone who worked for me, but they defeated me, and I had to resort to calling each of them "Bert" myself. Generally, we hired out the gangs to different clients and I had little to do with the work that they did. However, I got caught out one weekend when I was "volunteered" by Pino to look after a single line loose sleeper relaying job on the Leicester-Burton Line. That should have been a doddle for a former Renewals Manager in the prestigious East Coast Main Line right! Not a bit of it! It showed me how my Permanent Way Career had basically been about being able to organise to have the people who really knew what they were doing in the right place at the right time.

On this job, I had a supervisor (not a Bridgeway man) who had little interest in anything other than staying warm in his van, a "rookie" relaying gang and some on-track plant. The taking out of the old track had been completed eventually and the old ballast had been excavated but by the time that had been done, we were running several hours late. In my former existence, I just had to tell the relaying supervisor what I wanted and it got done. Now I found I was doing everything myself and not very well. The high point in ineptitude came when we started to relay the track. Now the key to a successful loose sleeper relay is to line up and space the sleepers in approximately the correct position for both line and level BEFORE positioning the rails, that way, the rails fit into the baseplates snugly with just some minor re-alignment of the sleepers as the work progresses. Not so with this team. In the rush to get the job done, they just dropped the sleepers randomly and then attempted to line them up once the rails were put on top of them. It was a complete disaster and I found that, without a good supervisor to pass on my instructions, I was unable to get it done correctly. We got less than half the work done that night before we gave up and I arranged for Karl to put his best supervisor on the job next time to finish the work while I stayed well clear. It was a chastening experience.

Near to Nottingham lies the Old Dalby Test Track on what used to be the Midland Main Line direct route from London to Nottingham. It extends from Asfordby in the South (where there is a connection to the national network) to Edwalton at the North end. The test track had been established in BR days and for most of the time that I was involved with it, it was operated by SERCO and used for testing of high-speed tilting trains and new London Underground stock. We were well-placed to win work there and over the years we secured a whole range of contracts. Before London Underground trains could be run, our track gangs installed the conductor rails, and we carried out a wide range of other trackwork. On the structures side, I examined tunnels, and our staff examined the bridges. We looked after their bridge-bash on-call and carried out structural repairs.

One day, with a colleague, I had to walk the line from Asfordby Depot through to Old Dalby but after all these years, I can't remember why. However, we signed in at Asfordby, received a briefing about train movements and off we set. It was a walk of about 5 miles, and we had to pass through 3 tunnels. The line was not built until the first decade of the 20th century and that showed in the quality of the work which was excellent. Through the cuttings between tunnels, there was a brick-lined channel to carry the drainage water rather than the more usual ditch and while we were approaching one of these, it became clear that there was a commotion up ahead. It became obvious that a large bird-of-prey which we identified as a buzzard had become trapped in the drainage channel and each time the bird tried to fly out, its wings struck the brick sides of the channel, and it was unable to take-off. We also noticed, some distance away, the reason why the buzzard was there; it had chased a small rabbit into the channel. The rabbit was cowering in the bottom of the channel, also unable to get out. Ideas of how to help were exchanged before I hit on a scheme. The sides of the cutting were wooded, and I found a large fallen branch which we placed in with one end in the bottom of the channel and the other on the bank above. We then retired a short distance away and waited. After a few minutes, the buzzard made its way along the channel to the branch and simply walked up it to the bank and took off. That was the first part of the rescue mission completed and we didn't have long to wait until the little rabbit also made its way up the branch and hopped off. That made our day, and we carried on walking along the railway with renewed enthusiasm.

Over the years, I had a hand in the provision of specialists of one form or another to the rail industry. Structures examination was a case in point. To carry out our diving inspections, we had to train and qualify our divers as bridge examiners. Not as easy as it sounds because in the early days of privatisation, there were few arrangements in place for staff training and companies instead were competing for a dwindling pool of ex-BR staff. Into this void we built a training organisation for our own needs and then expanded it into a commercial business training staff for other companies and Railtrack themselves. To provide bridge examiners, I designed a full course which included several weeks of both classroom and practical training. We got the course approved and I taught it for several years before getting additional trainers qualified and approved. Being able to train bridge examiners meant we could look to supply the industry with a range of specialist examiners including roped access and confined space. I found that I had quite an aptitude for teaching, and I really enjoyed the courses that I ran, although I did find it quite exhausting. Many of the examiners that I trained went on to get senior positions with Network Rail.

In the early days of Bridgeway, we got a request from Railtrack to arrange for the repair of damaged cast iron columns at York Station. Quite a lot of them had the decorative capitals damaged and because of the difficulty of repairing cast iron, replacements had to be designed, resourced, and fitted. For this, we contacted an industrial archaeologist, Ron Fitzgerald. Ron was an incredible draughtsman and designer and a knowledgeable historian. He was also a perfectionist, an idealist and somewhat eccentric. He made it clear to us from the start that if he was to be involved in the project, then only a perfect reproduction of the original capitols would be acceptable. They had to be cast iron and each one would have to be perfectly cast and painted. We presented him to Railtrack and, given his pedigree, they readily accepted him even though, when a Railtrack engineer said "wouldn't it be cheaper to use fibreglass" he was incredibly rude to our client. The drawings that Ron produced, all drawn by hand, were amazing and these were sent to a specialist foundry for casting. Needless to say, the first few castings didn't meet with Ron's approval but eventually, we got a good product which we fitted to the columns of the station. However, during the course of the job I found myself "piggy in the middle" between Ron and Railtrack but instead of the usual situation where contractor was trying to persuade the client to accept a slightly sub-standard product, Railtrack were begging me to stop Ron from delaying the completion of the job by insisting on perfection when they were quite happy with a few minor defects. In fact, at one stage, Ron's contemptuous attitude to the Railtrack Engineers was beginning to worry me in case it affected our being invited to bid for future work. I spent a lot of time smoothing ruffled feathers on both sides.

Ron Fitzgerald had many contacts and when we had need of surveyors, he introduced us to Mick and Paul. We had been given a small contract by Railtrack to survey both the position (relative to the railway above) and the condition of a brick culvert that meandered its way under Stafford Station. One of the platforms was subsiding and it was thought that a defective culvert might be to blame. The plan was therefore put forward and a budget item raised to line the culvert throughout in order restore its strength. Since the culvert passed under station buildings, platforms and 8 running lines, it was a major project but apparently, no one had ever been inside the culvert to look at the condition from within and there were no plans to show which route it followed. I dare say that the plans had existed at one time but probably finished up in the bins with many other BR archives. Examining the culvert was no problem for our confined space team (led, on this occasion, by Pino) but a complete survey required good surveyors – who could we get to go in the culvert with us?

Cue Mick and Paul. Both had been mine surveyors with the National Coal Board and had formed a small company of their own. Ron Fitzgerald introduced them to us, and it was the start of a partnership that lasted many years. In fact, in the end, Mick and Paul decided that it would be better to become employees of Bridgeway rather than sub-contractors and as such were part of a lot of major survey jobs. Well used to working in confined spaces in the mines, they shortened the legs of the tripod that carried the theodolite and followed the examiners through the culvert from end to end then repeated the exercise on the surface, thus accurately locating the route of the culvert on a plan of the station area. A good job all round and we thought that the Railtrack engineer would be delighted. He was, until Pino told him that the culvert was in very good condition throughout. His face dropped when he realised that all the preparations for the culvert had been in vain. The work was not needed and whatever had caused the platform to sink, it wasn't the culvert. I really liked working with Mick and Paul, they always had a ready solution for any problem, such as the time that we had to do a very complicated bed survey on the River Medway and they tracked the position of the boat with their instruments while it took sonar readings of the bed levels.

Anyway, that's enough of my time with Bridgeway, all of which I found immensely enjoyable and – let's be honest – more financially rewarding than working for a Nationalised Industry. I could continue but the story now belongs to Pino and his colleagues because the company continues to flourish and recently won the Queen's Award for Industry. However, I would say that, if privatisation had not happened, I would have been very happy to have followed my career with British Rail into retirement. In fact, in the 50 years of working in the Rail Industry, there were only 2 short periods when

I did not thoroughly enjoy it. Neither of these lasted more than 6 months so I count myself very lucky. A lot of people go to work every day without ever getting satisfaction from the job that they do. So, I retired completely in 2014 – or did I?

As I approached completion of this book in early 2022 at the age of 75, I was persuaded by Pino to give them a helping hand with a major bridge examination contract, so I had another 12-month spell of work before finally giving up in 2023.

Two 75-ton former steam breakdown cranes now converted to diesel carry out a tandem lift at Bridge 131 on the East Coast Main Line. Photo from the collection of fellow railway engineer Mark Wheel

9 781835 632871